# LNAT Practice Papers

## Volume One

ISBN 978-1-912557-31-8



Published by *RAR Medical Services Limited*
www.uniadmissions.co.uk
info@uniadmissions.co.uk
Tel: 0208 068 0438

# LNAT Mock Papers

## 2 Full Papers & Solutions

Aiden Ang
Rohan Agarwal

## About the Authors

**Aiden** graduated from Peterhouse, Cambridge, with a First Class Honours Law degree and has tutored Oxbridge law applicants at *UniAdmissions* for two years.

Aiden scored in the **top 10% nationally in the LNAT** and is now a trainee solicitor at a top US firm in London. He has a keen interest in helping out students with application advice as he believes that students should get all the help they need in order to succeed in their applications. In his spare time, he likes to travel and run outdoors.

**Rohan** is the **Director of Operations** at *UniAdmissions* and is responsible for its technical and commercial arms. He graduated from Gonville and Caius College, Cambridge and is a fully qualified doctor. Over the last five years, he has tutored hundreds of successful Oxbridge and Medical applicants. He has also authored ten books on admissions tests and interviews.

Rohan has taught physiology to undergraduates and interviewed medical school applicants for Cambridge. He has published research on bone physiology and writes education articles for the Independent and Huffington Post. In his spare time, Rohan enjoys playing the piano and table tennis.

# Introduction

### The Basics

The Law National Aptitude Test (LNAT) is a 2 hour 15 minutes test that is split into two sections – Section A comprises 42 multiple choice questions based on several passages, and Section B consists of a selection of essay questions from which you will have to attempt one.

Many top law schools use the LNAT, including University of Oxford, University College London and King's College London, hence it is imperative to do well for this test in order to maximise your chances of securing a spot in a top law school.

The only way to improve your LNAT scores, especially in Section A, is to keep practicing and reviewing your answers and examination technique. Hence, we have compiled a few LNAT Mock Papers that have been meticulously written by our expert tutors, designed to resemble the actual test as much as possible.

There is a dearth of information available freely which understandably makes students nervous and unprepared for the test. Our Mock Papers come with expert solutions that aim to let you know where you've gone wrong and prepare you as much as possible for the actual test.

### Preparing for the LNAT

Before going any further, it's important that you understand the optimal way to prepare for the LNAT. Rather than jumping straight into doing mock papers, it's essential that you start by understanding the components and the theory behind the LNAT by using a LNAT textbook. Once you've finished the non-timed practice questions, you can progress to using official LNAT papers. These are freely available online at **www.uniadmissions.co.uk/LNAT-past-papers** and serve as excellent practice. Finally, once you've exhausted past papers, move onto the mock papers in this book.

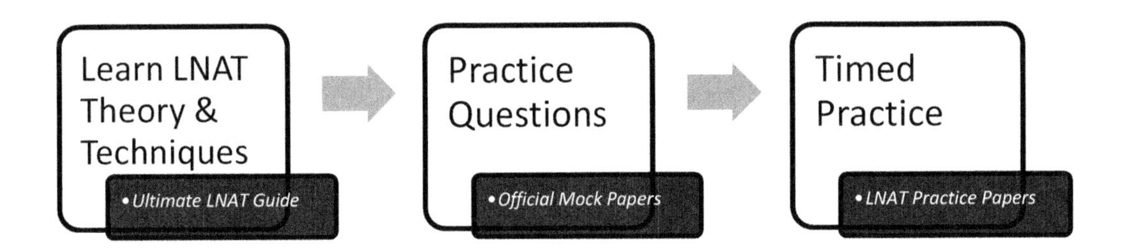


### Already seen them all?

So, you've run out of past papers? Well hopefully that is where this book comes in. It contains two unique mock papers; each compiled by Oxbridge law tutors at *UniAdmissions* and available nowhere else.

Having successfully gained a place on their course of choice, our tutors are intimately familiar with the LNAT and its associated admission procedures. So, the novel questions presented to you here are of the correct style and difficulty to continue your revision and stretch you to meet the demands of the LNAT.

## General Advice

### Start Early

It is much easier to prepare if you practice little and often. Start your preparation well in advance; ideally ten weeks but at the latest within a month. This way you will have plenty of time to complete as many papers as you wish to feel comfortable and won't have to panic and cram just before the test, which is a much less effective and more stressful way to learn. In general, an early start will give you the opportunity to identify the complex issues and work at your own pace.

### Prioritise

The MCQ section can be very time-pressured, and if you fail to answer the questions within the time limit you will be doing yourself a major disservice as every mark counts for this section. You need to be aware of how much time you're spending on each passage and allocate your time wisely. For example, since there are 42 questions in Section A, and you are given 95 minutes in total, you will ideally take about 130 seconds (just over two minutes) per question (including reading time for the passages) so that you will not run out of time and panic towards the end.

### Positive Marking

There are no penalties for incorrect answers; you will gain one for each right answer and will not get one for each wrong or unanswered one. This provides you with the luxury that you can always guess should you absolutely be not able to figure out the right answer for a question or run behind time. Since each question in Section A provides you with 4 possible answers, you have a 25% chance of guessing correctly. Therefore, if you aren't sure (and are running short of time), then make an educated guess and move on. Before 'guessing' you should try to eliminate a couple of answers to increase your chances of getting the question correct. For example, if a question has 4 options and you manage to eliminate 2 options- your chances of getting the question increase from 25% to 50%!

Avoid losing easy marks on other questions because of poor exam technique. Similarly, if you have failed to finish the exam, take the last 10 seconds to guess the remaining questions to at least give yourself a chance of getting them right.

### Practice

This is the best way of familiarising yourself with the style of questions and the timing for this section. Although the test does not demand any prior legal knowledge, you are unlikely to be familiar with the style of questions in all sections when you first encounter them. Therefore, you want to be comfortable at using this before you sit the test.

Practising questions will put you at ease and make you more comfortable with the exam. The more comfortable you are, the less you will panic on the test day and the more likely you are to score highly. Initially, work through the questions at your own pace, and spend time carefully reading the questions and looking at any additional data. When it becomes closer to the test, **make sure you practice the questions under exam conditions**.

### Past Papers

Official mock papers are freely available online at **www.uniadmissions.co.uk/lnat-past-papers**. Practice makes perfect, and the more you practice the questions, especially for Section A, the better you will get. Do not worry if you make plenty of mistakes at the start, the best way to learn is to understand why you have made certain mistakes and to not commit them again in the future!

### Repeat Questions

When checking through answers, pay particular attention to questions you have got wrong. If there is a worked answer, look through that carefully until you feel confident that you understand the reasoning, and then repeat the question without help to check that you can do it. If only the answer is given, have another look at the question and try to work out why that answer is correct. This is the best way to learn from your mistakes, and means you are less likely to make similar mistakes when it comes to the test. The same applies for questions which you were unsure of and made an educated guess which was correct, even if you got it right. When working through this book, **make sure you highlight any questions you are unsure of**, this means you know to spend more time looking over them once marked.

### No Dictionaries

The LNAT requires a strong command of the English language, especially for Section B where you are asked to write an essay in 40 minutes. You are not allowed to use spell check or a dictionary, hence you should ensure that you written English is up to standard and you should ideally make close to no grammatical or spelling errors for your essay.

Section A might also contain several passages that might be quite dense to the ordinary reader, but the LNAT is aimed at testing a student's reading comprehension skills. If there is a word you are unsure about, most of them time you should be able to deduce the meaning based on the context of the passage.

> ***Top tip!*** In general, universities tend to focus more on Section A of the LNAT in short listing candidates. Section B tends to be more subjective, and may be used simply as a tiebreaker at times.

### Keywords

If you're stuck on a question, sometimes you can simply quickly scan the passage for any keywords that match the questions. For example, by searching for 'old' in the passage or words related to 'old', it will help you to answer the question.

### A word on timing...

**"If you had all day to do your exam, you would get 100%. But you don't."**

Whilst this isn't completely true, it illustrates a very important point. Once you've practiced and know how to answer the questions, the clock is your biggest enemy. This seemingly obvious statement has one very important consequence. **The way to improve your score is to improve your speed.** There is no magic bullet. But there are a great number of techniques that, with practice, will give you significant time gains, allowing you to answer more questions and score more marks.

Timing is tight throughout – **mastering timing is the first key to success**. Some candidates choose to work as quickly as possible to save up time at the end to check back, but this is generally not the best way to do it. Often questions can have a lot of information in them – each time you start answering a question it takes time to get familiar with the instructions and information. By splitting the question into two sessions (the first run-through and the return-to-check) you double the amount of time you spend on familiarising yourself with the data, as you have to do it twice instead of only once. This costs valuable time. In addition, candidates who do check back may spend 2–3 minutes doing so and yet not make any actual changes. Whilst this can be reassuring, it is a false reassurance as it is unlikely to have a significant effect on your actual score. Therefore, it is usually best to pace yourself very steadily, aiming to spend the same amount of time on each question and finish the final question in a section just as time runs out. This reduces the time spent on re-familiarising with questions and maximises the time spent on the first attempt, gaining more marks.

**It is essential that you don't get stuck with the hardest questions** – no doubt there will be some. In the time spent answering only one of these you may miss out on answering three easier questions. If a question is taking too long, choose a sensible answer and move on. Never see this as giving up or in any way failing, rather it is the smart way to approach a test with a tight time limit. With practice and discipline, you can get very good at this and learn to maximise your efficiency. It is not about being a hero and aiming for full marks – this is almost impossible and very much unnecessary (even Oxford will regard any score higher than 30 out of 42 as exceptional). It is about maximising your efficiency and gaining the maximum possible number of marks within the time you have.

### Manage your Time:

It is highly likely that you will be juggling your revision alongside your normal school studies. Whilst it is tempting to put your A-levels on the back burner falling behind in your school subjects is not a good idea, don't forget that to meet the conditions of your offer should you get one you will need at least one A*. So, time management is key!

Make sure you set aside a dedicated 90 minutes (and much more closer to the exam) to commit to your revision each day. The key here is not to sacrifice too many of your extracurricular activities, everybody needs some down time, but instead to be efficient. Take a look at our list of top tips for increasing revision efficiency below:

1. Create a comfortable work station
2. Declutter and stay tidy
3. Treat yourself to some nice stationery
4. See if music works for you → if not, find somewhere peaceful and quiet to work
5. Turn off your mobile or at least put it into silent mode; silence social media alerts
6. Keep the TV off and out of sight
7. Stay organised with to do lists and revision timetables – more importantly, stick to them!
8. Keep to your set study times and don't bite off more than you can chew
9. Study while you're commuting
10. Adopt a positive mental attitude and get into a routine
11. Consider forming a study group to focus on the harder exam concepts
12. Plan rest and reward days into your timetable – these are excellent incentive for you to stay on track with your study plans!

**Use the Options:**

Some passages may try to trick you by providing a lot of unnecessary information. When presented with long passages that are seemingly hard to understand, it's essential you look at the answer options so you can focus your mind. This can allow you to reach the correct answer a lot more quickly. Consider the example below:

*What, then, is it that gives to Sanskrit its claim on our attention, and its supreme importance in the eyes of the historian? First of all, its antiquity—for we know Sanskrit at an earlier period than Greek. But what is far more important than its merely chronological antiquity is the antique state of preservation in which that Aryan language has been handed down to us. The world had known Latin and Greek for centuries, and it was felt, no doubt, that there was some kind of similarity between the two. But how was that similarity to be explained? Sometimes Latin was supposed to give the key to the formation of a Greek word, sometimes Greek seemed to betray the secret of the origin of a Latin word. Afterward, when the ancient Teutonic languages, such as Gothic and Anglo-Saxon, and the ancient Celtic and Slavonic languages too, came to be studied, no one could help seeing a certain family likeness among them all. But how such a likeness between these languages came to be, and how, what is far more difficult to explain, such striking differences too between these languages came to be, remained a mystery, and gave rise to the most gratuitous theories, most of them, as you know, devoid of all scientific foundation. As soon, however, as Sanskrit stepped into the midst of these languages, there came light and warmth and mutual recognition. They all ceased to be strangers, and each fell of its own accord into its right place. Sanskrit was the eldest sister of them all, and could tell of many things which the other members of the family had quite forgotten. Still, the other languages too had each their own tale to tell; and it is out of all their tales together that a chapter in the human mind has been put together which, in some respects, is more important to us than any of the other chapters, the Jewish, the Greek, the Latin, or the Saxon. The process by which that ancient chapter of history was recovered is very simple. Take the words which occur in the same form and with the same meaning in all the seven branches of the Aryan family, and you have in them the most genuine and trustworthy records in which to read the thoughts of our true ancestors, before they had become Hindus, or Persians, or Greeks, or Romans, or Celts, or Teutons, or Slaves. Of course, some of these ancient charters may have been lost in one or other of these seven branches of the Aryan family, but even then, if they are found in six, or five, or four, or three, or even two only of its original branches, the probability remains, unless we can prove a later historical contact between these languages, that these words existed before the great Aryan Separation. If we find agni, meaning fire, in Sanskrit, and ignis, meaning fire, in Latin, we may safely conclude that fire was known to the undivided Aryans, even if no trace of the same name of fire occurred anywhere else. And why? Because there is no indication that Latin remained longer united with Sanskrit than any of the other Aryan languages, or that Latin could have borrowed such a word from Sanskrit, after these two languages had once become distinct. We have, however, the Lithuanian ugnìs, and the Scottish ingle, to show that the Slavonic and possibly the Teutonic languages also, knew the same word for fire, though they replaced it in time by other words. Words, like all other things, will die, and why they should live on in one soil and wither away and perish in another, is not always easy to say. What has become of ignis, for instance, in all the Romance languages? It has withered away and perished, probably because, after losing its final unaccentuated syllable, it became awkward to pronounce; and another word, focus, which in Latin meant fireplace, hearth, altar, has taken its place.*

This is an extremely dense passage with a lot of information. **Looking at the options first makes it obvious that certain information are redundant** and allows you to quickly zoom in on certain keywords you should pick up on in order to answer the questions.

In other cases, **you may actually be able to solve the question without having to read the passage over and over again**. For example:

Which language does the writer state is the oldest?
A. Sanskrit B. Greek C. Latin D. Gothic E. Anglo-Saxon

If you read the passage first before looking at the question, you might have forgotten what the passage mentioned about which language was the oldest, and you will have to spend extra time going back to the passage to re-read it again.

You can save a lot of time by looking at the questions first before reading the passage. After looking at the question, you will know at the back of your head to look out for which language was stated as the oldest by the author, and this will save a considerable amount of time.

## Keep Fit & Eat Well:

*'A car won't work if you fill it with the wrong fuel'* - your body is exactly the same. You cannot hope to perform unless you remain fit and well. The best way to do this is not underestimate the importance of healthy eating. Beige, starchy foods will make you sluggish; instead start the day with a hearty breakfast like porridge. Aim for the recommended 'five a day' intake of fruit/veg and stock up on the oily fish or blueberries – the so called "super foods".

When hitting the books, it's essential to keep your brain hydrated. If you get dehydrated you'll find yourself lethargic and possibly developing a headache, neither of which will do any favours for your revision. Invest in a good water bottle that you know the total volume of and keep sipping throughout the day. Don't forget that the amount of water you should be aiming to drink varies depending on your mass, so calculate your own personal recommended intake as follows: 30 ml per kg per day.

It is well known that exercise boosts your wellbeing and instils a sense of discipline. All of which will reflect well in your revision. It's well worth devoting half an hour a day to some exercise, get your heart rate up, break a sweat, and get those endorphins flowing.

## Sleep

It's no secret that when revising you need to keep well rested. Don't be tempted to stay up late revising as sleep actually plays an important part in consolidating long term memory. Instead aim for a minimum of seven hours good sleep each night, in a dark room without any glow from electronic appliances. Install flux (https://justgetflux.com) on your laptop to prevent your computer from disrupting your circadian rhythm. Aim to go to bed the same time each night and no hitting snooze on the alarm clock in the morning!

## Revision Timetable

Still struggling to get organised? Then try filling in the example revision timetable below, remember to factor in enough time for short breaks, and stick to it! Remember to schedule in several breaks throughout the day and actually use them to do something you enjoy e.g. TV, reading, YouTube etc.

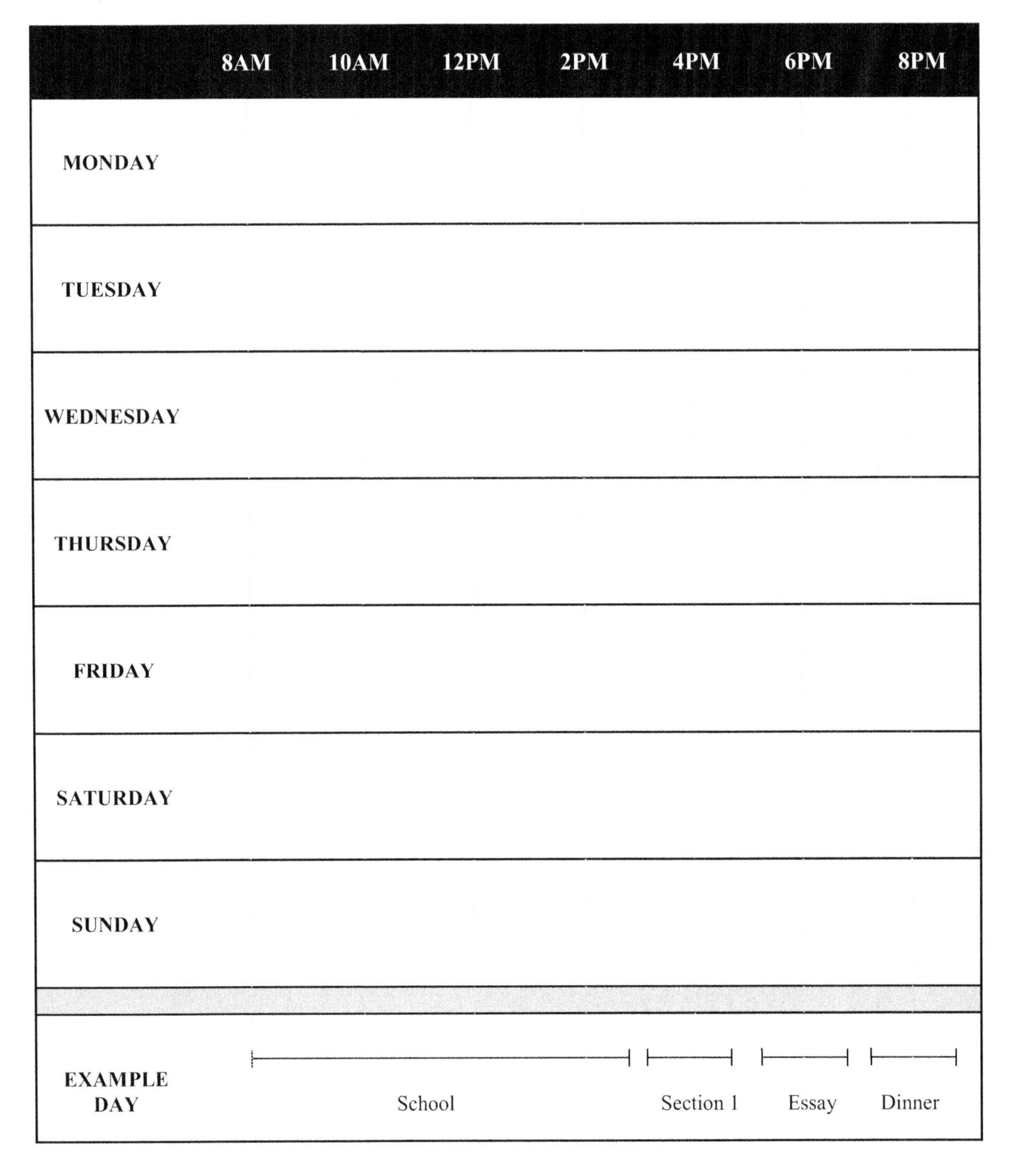

| | 8AM | 10AM | 12PM | 2PM | 4PM | 6PM | 8PM |
|---|---|---|---|---|---|---|---|
| MONDAY | | | | | | | |
| TUESDAY | | | | | | | |
| WEDNESDAY | | | | | | | |
| THURSDAY | | | | | | | |
| FRIDAY | | | | | | | |
| SATURDAY | | | | | | | |
| SUNDAY | | | | | | | |
| EXAMPLE DAY | | | School | | Section 1 | Essay | Dinner |

***Top tip!*** Ensure that you take a watch that can show you the time in seconds into the exam. This will allow you have a much more accurate idea of the time you're spending on a question. In general, if you've spent >180 seconds on a section 1 question – move on regardless of how close you think you are to solving it.

## Getting the most out of Mock Papers

Mock exams can prove invaluable if tackled correctly. Not only do they encourage you to start revision earlier, they also allow you to **practice and perfect your revision technique**. They are often the best way of improving your knowledge base or reinforcing what you have learnt. Probably the best reason for attempting mock papers is to familiarise yourself with the exam conditions of the LNAT as they are particularly tough.

### Start Revision Earlier

Thirty five percent of students agree that they procrastinate to a degree that is detrimental to their exam performance. This is partly explained by the fact that they often seem a long way in the future. In the scientific literature this is well recognised, Dr. Piers Steel, an expert on the field of motivation states that *'the further away an event is, the less impact it has on your decisions'*.

Mock exams are therefore a way of giving you a target to work towards and motivate you in the run up to the real thing – every time you do one treat it as the real deal! If you do well then it's a reassuring sign; if you do poorly then it will motivate you to work harder (and earlier!).

### Practice and perfect revision techniques

In case you haven't realised already, revision is a skill all to itself, and can take some time to learn. For example, the most common revision techniques including **highlighting and/or re-reading are quite ineffective** ways of committing things to memory. Unless you are thinking critically about something you are much less likely to remember it or indeed understand it.

Mock exams, therefore allow you to test your revision strategies as you go along. Try spacing out your revision sessions so you have time to forget what you have learnt in-between. This may sound counterintuitive but the second time you remember it for longer. Try teaching another student what you have learnt; this forces you to structure the information in a logical way that may aid memory. Always try to question what you have learnt and appraise its validity. Not only does this aid memory but it is also a useful skill for the LNAT, Oxbridge interviews, and beyond.

### Improve your knowledge

The act of applying what you have learnt reinforces that piece of knowledge. An essay question may ask you about a fairly simple topic, but if you have a deep understanding of it you are able to write a critical essay that stands out from the crowd. Essay questions in particular provide a lot of room for students who have done their research to stand out, hence you should always aim to improve your knowledge and apply it from time to time. As you go through the mocks or past papers take note of your performance and see if you consistently under-perform in specific areas, thus highlighting areas for future study.

### Get familiar with exam conditions

Pressure can cause all sorts of trouble for even the most brilliant students. The LNAT is a particularly time pressured exam with high stakes – your future (without exaggerating) does depend on your result to a great extent. The real key to the LNAT is overcoming this pressure and remaining calm to allow you to think efficiently.

Mock exams are therefore an excellent opportunity to devise and perfect your own exam techniques to beat the pressure and meet the demands of the exam. **Don't treat mock exams like practice questions – it's imperative you do them under time conditions.**

***Remember!*** It's better that you make all the mistakes you possibly can now in mock papers and then learn from them so as not to repeat them in the real exam.

## Before using this Book

### Do the ground work

- Understand the format of the LNAT – have a look at the LNAT website and familiarise yourself with it: www.lnat.ac.uk/test-format
- Improve your written English by practicing writing and reading frequently.
- Try to broaden your reading by learning about different topics that you are unfamiliar with as the essay topics can vary greatly.
- Learn how to understand a writer's viewpoint by reading news articles and having a go at summarising what the writer is arguing about.
- Be consistent – slot in regular LNAT practice sessions when you have pockets of free time.
- Engage in discussion sessions with your friends to give you more ideas about certain essay topics.
- Download the LNAT simulator and have a go at doing in online – the actual test is done on a computer so you will want to be familiar with the format – www.uniadmissions.co.uk/lnat-past-papers

### Ease in gently

With the ground work laid, there's still no point in adopting exam conditions straight away. Instead invest in a beginner's guide to the LNAT, which will not only describe in detail the background and theory of the exam, but take you through section by section what is expected. *The Ultimate LNAT Guide* is the most popular LNAT textbook – you can get a free copy by flicking to the back of this book.

Questions are seldom repeated, so don't rote learn methods or facts. Instead, focus on applying prior knowledge to formulate your own approach. If you're really struggling and have to take a sneak peek at the answers, then practice thinking of alternative solutions, or arguments for essays. It is unlikely that your answer will be more elegant or succinct than the model answer, but it is still a good task for encouraging creativity with your thinking. Get used to thinking outside the box!

### Accelerate and Intensify

Start adopting exam conditions after you've done the official mock papers. Remember that **time pressure makes the LNAT hard** – if you had as long as you wanted to sit the exam you would probably get 100%.

Doing all the mock papers in this book is a good target for your revision. Choose a paper and proceed with strict exam conditions. Take a short break and then mark your answers before reviewing your progress. For revision purposes, as you go along, keep track of those questions that you guess – these are equally as important to review as those you get wrong.

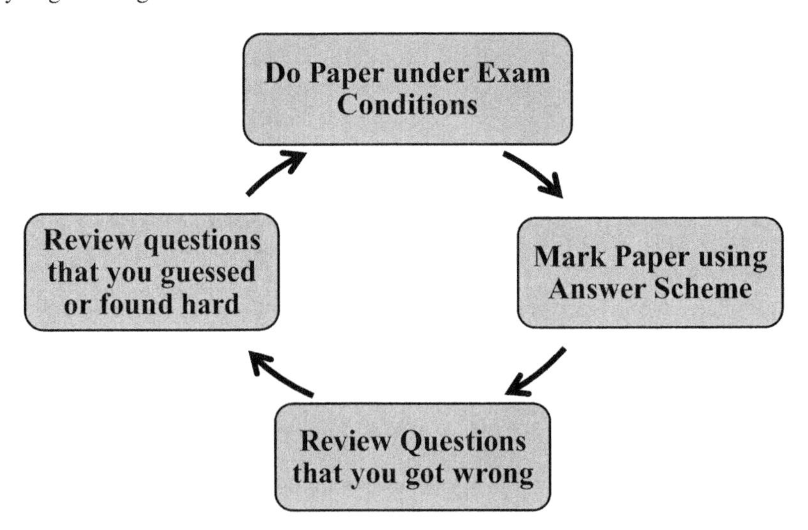


Once you've exhausted all the past papers, move on to tackling the unique mock papers in this book. In general, you should aim to complete mock papers every night in the seven days preceding your exam.

## Section A: An Overview

| What will you be tested on? | No. of Questions | Duration |
|---|---|---|
| Reading comprehension skills, deducing arguments, understanding certain literary tools, questioning assumptions | 42 MCQs | 95 Minutes |

This is the first section of the LNAT, comprising several passages to read and a total of 42 MCQ questions. You have 95 minutes in total to complete the MCQ questions, including reading time for 10-12 passages. In order to keep within the time limit, you realistically have about two minutes per question as you have to factor in the reading time for the passages as well.

Not all the questions are of equal difficulty and so as you work through the past material it is certainly worth learning to recognise quickly which questions you should spend less time on in order to give yourself more time for the trickier questions.

### Deducing arguments

Several MCQ questions will be aimed at testing your understand of the writer's argument. It is common to see questions asking you 'what is the writer's view?' or 'what is the writer trying to argue?'. This is arguably an important skill you will have to develop as a law student, and the LNAT is designed to test this ability. You have limited time to read the passage and understand the writer's argument, and the only way to improve your reading comprehension skill is to read several well-written news articles on a daily basis and think about them in a critical manner.

### Assumptions

It is important to be able to identify the assumptions that a writer makes in the passage, as several questions might question your understand of what is assumed in the passage. For example, if a writer mentions that 'if all else remains the same, we can expect our economic growth to improve next year', you can identify an assumption being made here – the writer is clearly assuming that all external factors remain the same.

### Fact vs. Opinion

It is important to be able to decipher whether the writer is stating a fact or an opinion – the distinction is usually rather subtle and you will have to decide whether the writer is giving his or her own personal opinion, or presenting something as a fact. Section A may contain questions that will test your ability to identify what is presented as a fact and what is presented as an opinion.

| Fact | Opinion |
|---|---|
| 'There are 7 billion people in this world...' | 'I believe there are more than 7 billion people in this world...' |
| 'She is an Australian...' | 'She sounded like an Australian...' |
| 'Trump is the current President...' | 'Trump is a horrible President...' |
| 'Vegetables contain a lot of fibre...' | 'Vegetables are good for you...' |

## Section B: An Overview

| What will you be tested on? | No. of Questions | Duration |
|---|---|---|
| Your ability to write an essay under timed conditions, your writing technique and your argumentative abilities | 1 out of 3 | 40 Minutes |

Section 2 is usually what students are more comfortable with – after all, many GCSE and A Level subjects require you to write essays within timed conditions. It does not require you to have any particular legal knowledge – the questions can be very broad and cover a wide range of topics.

Here are some of the topics that might appear in Section B:

- Science
- Politics
- Religion
- Technology
- Ethics
- Morality
- Philosophy
- Education
- History
- Geopolitics

As you can see, this list is very broad and definitely non-exhaustive, and you do not get many choices to choose from (you have to write one essay out of three choices). Many students make the mistake of focusing too narrowly on one or two topics that they are comfortable with – this is a dangerous gamble and if you end up with three questions you are unfamiliar with, this is likely to negatively impact your score.

You should ideally focus on 3-4 topics to prepare from the LNAT, and you can pick and choose which topics from the list above are the ones you would be more interested in. Here are some suggestions:

**Science**
An essay that is related to science might relate to recent technological advancements and their implications, such as the rise of Bitcoin and the use of blockchain technology and artificial intelligence. This is interrelated to ethical and moral issues, hence you cannot merely just regurgitate what you know about artificial intelligence or blockchain technology. The examiners do not expect you to be an expert in an area of science – what they want to see is how you identify certain moral or ethical issues that might arise due to scientific advancements, and how do we resolve such conundrums as human beings.

**Politics**
Politics is undeniably always a hot topic and consequently a popular choice amongst students. The danger with writing a politics question is that some students get carried away and make their essay too one-sided or emotive – for example a student may chance upon an essay question related to Brexit and go on a long rant about why the referendum was a bad idea. You should always remember to answer the question and make sure your essay addresses the exact question asked – do not get carried away and end up writing something irrelevant just because you have strong feelings about a certain topic.

**Religion**
Religion is always a thorny issue and essays on religion provide strong students with a good opportunity to stand out and display their maturity in thought. Questions can range from asking about your opinion with regards to banning the wearing of a headdress to whether children should be exposed to religious practices at a young age. Questions related to religion will require a student to be sensitive and measured in their answers and it is easy to trip up on such questions if a student is not careful.

**Education**
Education is perhaps always a relatable topic to students, and students can draw from their own experience with the education system in order to form their opinion and write good essays on such topics. Questions can range from whether university places should be reduced, to whether we should be focusing on learning the sciences as opposed to the arts.

## Section B: Revision Guide

### SCIENCE

| Resource | What to read/do |
|---|---|
| 1. **Newspaper Articles** | • The Guardian, The Times, The Economist, The Financial Times, The Telegraph, The New York Times, The Independent |
| 2. **A Levels/IB** | • Look at the content of your science A Levels/IB if you are doing science subjects and critically analyse what are the potential moral/ethical implications<br>• Use your A Levels/IB resources in order to seek out further readings – e.g. links to a scientific journal or blog commentary<br>• Remember that for your LNAT essay you should not focus on the technical issues too much – think more about the ethical and moral issues |
| 3. **Online videos** | • There are plenty of free resources online that provide interesting commentary on science and the moral and ethical conundrums that scientists face on a daily basis<br>• E.g. Documentaries and specialist science channels on YouTube<br>• National Geographic, Animal Planet etc. might also be good if you have access to them |
| 4. **Debates** | • Having a discussion with your friends about topics related to science might also help you formulate some ideas<br>• Attending debate sessions where the topic is related to science might also provide you with excellent arguments and counter-arguments<br>• Some universities might also host information sessions for sixth form students – some might be relevant to ethical and moral issues in science |
| 5. **Museums** | • Certain museums such as the Natural Science Museum might provide some interesting information that you might not have known about |
| 6. **Non-fiction books** | • There are plenty of non-fiction books (non-technical ones) that might discuss moral and ethical issues about science in an easily digestible way |

### POLITICS

| Resource | What to read/do |
|---|---|
| 1. **Newspaper Articles** | • The Guardian, The Times, The Economist, The Financial Times, The Telegraph, The New York Times, The Independent |
| 2. **Television** | • Parliamentary sessions<br>• Prime Minister Questions<br>• Political news |
| 3. **Online videos** | • Documentaries<br>• YouTube Channels |
| 4. **Lectures** | • University introductory lectures<br>• Sixth form information sessions |
| 5. **Debates** | • Debates held in school<br>• Joining a politics club |
| 6. **Podcasts** | • Political podcasts<br>• Listen to both sides to get a more rounded view (e.g. listening to both left and right wing podcasts) |

## RELIGION

| Syllabus Point | What to read/do |
|---|---|
| 1. **Newspaper Articles** | • The Guardian, The Times, The Economist, The Financial Times, The Telegraph, The New York Times, The Independent |
| 2. **Non-fiction books** | • Read up about books that explain the origins and beliefs of different types of religion<br>• E.g. Books that talk about the origins of Christianity, Islam or Buddhism, theology books etc. |
| 3. **Talking to religious leaders** | • Talking to religious leaders may be a good way of understanding different religions more and being able to write an essay on religion with more maturity and nuance<br>• Talking to people from different religious backgrounds may also be a good way of forming a more well-rounded opinion |
| 4. **Online videos** | • Documentaries on religion<br>• YouTube channels providing informative and educational videos on different religions – e.g. history, background |
| 5. **Lectures** | • Information sessions<br>• Relevant introductory lectures |
| 6. **Opinion articles** | • Informative blogs and journals<br>• Read both arguments and counter-arguments and come up with your own viewpoint |

## EDUCATION

| Syllabus Point | What to read/do |
|---|---|
| 1. **Newspaper Articles** | • The Guardian, The Times, The Economist, The Financial Times, The Telegraph, The New York Times, The Independent |
| 2. **A Levels/IB** | • Draw inspiration from what you are studying in your A Levels or IB – do you feel like what you are studying is useful and relevant? E.g. Studying arts versus science<br>• Compare the education you are receiving with your friends in different schools or different subjects |
| 3. **Educational exchange** | • If you have an opportunity to go on an educational exchange, this might be a good opportunity to compare and contrast different educational systems<br>• E.g. the approach to education in Germany versus the UK |
| 4. **University applications** | • Have a read of how different universities promote themselves – do they claim to provide students with academic enlightenment, or better job prospects, or a good social life?<br>• Why do different universities focus on different things? |
| 5. **Online videos** | • Documentaries<br>• YouTube Channels |
| 6. **Talk to your teachers** | • Your teachers have been in the education industry for years and maybe decades – talk to them and ask them for their opinion<br>• Talk to different teachers and compare their opinions regarding how we should approach education |

***Top Tip!*** Although you aren't required to have extra knowledge for the LNAT essay, doing so will allow you to make your essay stand out from the crowd. However, you should first **prioritise perfecting your writing style rather than doing extra reading** as the former will have a greater impact on your mark.

## How to use this Book

If you have done everything this book has described so far then you should be well equipped to meet the demands of the LNAT, and therefore **the mock papers in the rest of this book should ONLY be completed under exam conditions**. This means:

- Absolute silence – no TV or music
- Absolute focus – no distractions such as eating your dinner
- Strict time constraints – no pausing half way through
- No checking the answers as you go
- Give yourself a maximum of three minutes between sections – keep the pressure up
- Complete the entire paper before marking and mark harshly

This means setting aside 2 hours and 15 minutes to tackle the paper. Completing one mock paper every evening in the week running up to the exam is an ideal target.

- Return to mark your answers, but mark harshly if there's any ambiguity.
- Highlight any areas of concern and read up on the areas you felt you underperformed
- If you inadvertently learnt anything new by muddling through a question, go and tell somebody about it to reinforce what you've discovered.

Finally relax… the LNAT is an exhausting exam, concentrating so hard continually for two hours will take its toll. So, being able to relax and switch off is essential to keep yourself sharp for exam day! Make sure you reward yourself after you finish marking your exam.

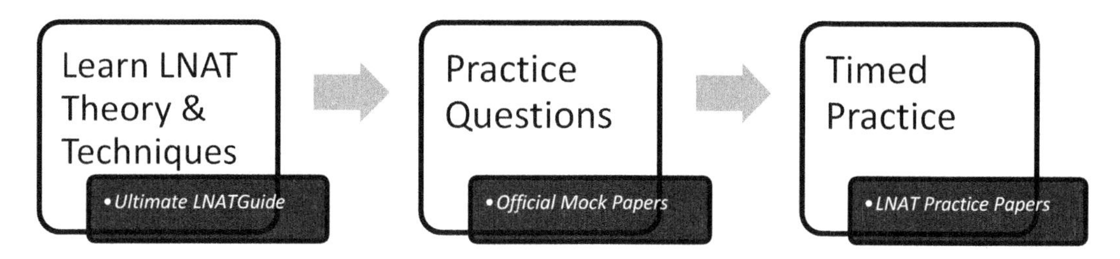


> ***Top Tip!*** You can get free copies of *The Ultimate LNAT Guide* by flicking to the back of this book.

## Scoring Tables

Use these to keep a record of your scores from past papers – you can then easily see which paper you should attempt next (always the one with the lowest score).

| | SECTION A | 1st Attempt | 2nd Attempt | 3rd Attempt |
|---|---|---|---|---|
| Volume One | Mock Paper A | | | |
| | Mock Paper B | | | |
| Volume Two | Mock Paper C | | | |
| | Mock Paper D | | | |

You will not be able to give yourself a score for Section 2 per se as an essay is always marked rather subjectively – the best way to gauge your performance for Section 2 will be to compare your arguments and counter-arguments with the model answer, or get feedback from your teachers.

> ***Remember!*** You can get a free copy of Volume 2 (Papers C and D) of LNAT *Practice Papers* by flicking to the back of this book.

# Mock Papers

## Mock Paper A

### 1. Disc or Download: A Virtual Energy-Savings Debate

Adapted from 2014©. Copyright Environmental News Network

One of the best ways to spark an energy revolution is through the younger generation — and nothing quite speaks their language like video games.

But this issue has less to do with the content of these addictive games and more with how the younger generation consumes them.

Fantasy and adventure, sci-fi and first-person shooters, strategy and racing — video games today come in all types of genres with thousands of add-ons and customizable features to make each story a virtual reality. And with all of these choices comes two more: buy a copy of the video game on a disc or download the video game straight from the console?

At the crossroads of the decision lies Sony, one of the leading manufacturers of video games including the PlayStation® console. Sony offers the choice between direct download through an Internet connection and purchasing via game stores a Blu-ray copy of the game.

Now, one would think the former option is friendlier on the environment — no carbon footprint from shipping or driving to stores. However, cartridge collectors everywhere are about to rejoice. According to the *Journal of Industrial Ecology*, downloading a game from an Internet server can create a larger carbon footprint than driving to a store to purchase the game on Blu-ray disc.

A study was conducted by the *Journal of Industrial Ecology*, and although the report was based on a series of assumptions, the results were as follows:

"The CF [carbon footprint] of the life cycle of a downloaded 8.80 GB game amounted to 21.9 to 27.5 kg $CO_2$-eq (for lower and upper bounds of Internet energy intensity), whereas the result for a BD [Blu-ray disc] game was 20.8 kg $CO_2$-eq. Gameplay (use phase) accounted for 19.5 kg $CO_2$-eq emissions in both scenarios."

As observed, the carbon footprint from driving to the store and purchasing a Blu-ray disc is less comparatively, but there are some caveats. There is a threshold of 1.3 gigabytes, under which PlayStation® games are still more efficient to download. PlayStation® games are only getting larger, though.

According to the *Journal of Industrial Ecology*, game files have doubled in size between 2010 and 2013, and have increased by 25 percent from PlayStation 3 to PlayStation 4.

In fact, the advantage of discs is their ability to store massive amounts of data, which allows them to become exponentially better than direct downloads as their GB capacity increases. All that's needed is a little more transparency for all video games to be rated "E" — no, not "E" for everyone, but "E" for efficient.

1. Why does the author compare '"E" for everyone' and '"E" for efficient?

A. To illustrate the importance of efficiency
B. To use humour to emphasise the importance of universality
C. To bring together themes of environmentalism and gaming
D. To show that efficiency is not the main concern of video-game makers
E. To argue that the efficiency of video games should be more visible to consumers

2. Which of the following is an assumption that the author has made?

A. The younger generation will be more likely to engage in environmental debate if it surrounds a topic which interests them
B. That transparency regarding the efficiency of video games will encourage younger people to think more carefully about which games they purchase
C. That a higher carbon footprint is more detrimental for the environment
D. That a significant proportion of the younger generation have an interest in video games
E. Discs care better than downloads as they are capable of being shared amongst friends

3. Why does the author use a question as a conclusion to the third paragraph?

A. He wants the reader to engage in the debate through online comment
B. He implies that the issue of efficiency should be as important as the choice between different types of game
C. He wants to show that young people are overwhelmed by questions
D. He attempts to show that there is a real choice to be made between the two options, one which shouldn't be overlooked
E. He is trying to create an informal tone to the piece

## 2. Why Homework Matters

Adapted from 2014©. Edarticle

As an elementary/middle school teacher, I hear constant complaints about the issue of homework. There are valid points against overdoing it and even studies that suggest, in some cases, it doesn't always help. There's a big difference between busy work and assignments that are meaningful. Some researchers, like Sara Bennett and Nancy Kalish, propose that homework is a hidden cause of childhood obesity. Others, like Alfie Kohn, believe that the quality and quantity of assignments done at home should be addressed, pronto. So, why do students today still have to do this archaic activity?

- Although some teachers assign busy work, that is not always the case. Often times, the assignments students bring home are not only continuations of what was done in class, but are activities that help apply the knowledge learned earlier in the day.
- Some at-home assignments are enrichment activities that will nurture the creative side of children.
- It allows parents to see at firsthand what their child is doing in school, and gives them an opportunity to connect or communicate with their child on a different level, seeing where they stand academically.
- Even if it's studying or reviewing what was covered in class, it is an important reinforcement activity. This can eventually lead to information being stored into long-term memory.
- Research conducted by Harris Cooper (2006) suggests that students who complete homework tend to have higher scores on content related tests, than do students who have not completed content related homework.
- It can teach a child responsibility, which in turn will form into a good work ethic that will be useful for the rest of his/her life.
- Time management skills will develop and students will learn a valuable lesson about procrastination.
- It will allow students the opportunity to teach themselves, and learn how to manage schoolwork independently.

Although there is an ongoing debate about the merits and necessity of homework, the bottom line is that it doesn't look like it's going away any time soon. Sure, parents should question teachers and make sure that the homework is relevant and not going to last five hours, but the best possible action a student can take at the elementary age is to accept its inevitability. This will lead to time management strategies, allowing them to learn how to juggle a variety of activities on a daily basis, just like parents and all other adults do every day.

Studies at Duke University under Harris Cooper found a positive correlation between test scores and homework. "The results of such studies suggest that homework can improve students' scores on the class tests that come at the end of a topic. Students assigned homework in 2nd grade did better in mathematics, 3rd and 4th graders did better on English skills and vocabulary, 5th graders on social studies, 9th through 12th graders on American history, and 12th graders on Shakespeare." Of course, when it is overloaded, there's the case of diminishing returns.

Life in the working world is more competitive than ever before. Children need to learn the methods that will allow them to be successful in the life that's just ahead. Kohn suggests that homework doesn't help elementary school students, when research states otherwise. He is also a proponent of eliminating grades. Kohn's ideas generate excellent debate topics, but are not practical in a world that uses rankings for everything. So, even if you don't buy into the fact that homework will make a child a higher academic achiever in the short term (even though research states otherwise), realize that it just might create a human being with good habits, a rich work ethic, and success later on in the world outside academia.

4. Which of the following is a statement of fact rather than an assertion of opinion?

A. 'Children need to learn the method that will allow them to be more successful'

B. 'Kohn's ideas generate excellent debate'

C. 'Some researchers, like Sara Bennett and Nancy Kalish propose that homework is a hidden cause of childhood obesity'

D. 'There's a big difference between busy work and assignments that are meaningful'

E. 'it just might create a human being with good habits'

5. Which of the following does the author imply in the piece

A. Nurturing of a child's personality is equally as important as academic success

B. Good habits will help a child throughout his/her adult life

C. That s/he is fed up with hearing complaints about the issue of homework

D. Teachers who assign busy work are failing their students

E. Kohn's research is not trustworthy

6. What, according to the author, do children need to learn?

A. The methods to help them earn money in their later life

B. The mindset required in order to become successful in adulthood

C. To manage their schoolwork independently

D. Techniques that will help them to be successful

E. How to deal with an overload of homework

7. What is the main conclusion of the article?

A. Homework leads to good grades, which is important for children's prospects

B. Too much homework can be overwhelming for children

C. Homework is not helpful for academic achievement, but is instrumental in creating a rounded human being

D. Parents' opinions on how much homework their children have should be taken into consideration

E. Homework is likely to be helpful in a child's development, however important it ends up being from an academic perspective

### 3. What would independence mean for Scotland's racial minorities?

Adapted from 2014© Nasar Meer for the Guardian

*Scotland's 'other' has firmly been the English. What happens when it starts to look more inward?*

On Edinburgh's Royal Mile, the ancient road running from its 12th-century castle at the top down to the Enric Miralles-designed parliament at its bottom, you'll find some unlikely proponents of Scottish nationalism. Sporting tartan turbans and proudly brandishing the Saltire, the Sikh small-shop owners are sometimes viewed curiously by tourists and festival goers. The example is symbolic because at a time when Scottish identity is being appropriated in various arenas, it does raise the question of where ethnic and racial minorities fit into a country dominated by myths and legends of an ostensibly "white" nation dating back millennia.

Another way of putting this is to say that whatever else "Britishness" might be, we know that it's not the sole preserve of white Christians, but we're less sure about "Scottishness" (as we might be of "Englishness" if we looked close enough). Of course it depends where you go in Scotland, as it does in Britain as a whole, but surveys tell us that the Scots are no more exclusionary in their attitudes than the English.

With only around 2% ethnic minorities, however, it's a theory of tolerance that is yet to be tested and, as any student of nationalism will tell you, there's a fine line between inclusive and exclusive national identities. Until now though, and sectarian issues aside, it appears that Scotland's "other" has firmly been the English – what happens when it starts to look more inward?

Perhaps with the exception of Herman Rodrigues' 2006 exhibition on Scotland's Asian communities, little is said of the "new Scots". It is nevertheless a matter of enormous pride to the SNP that the only ethnic minority MSPs have been members of their party, the most recent being Humza Yousaf who earlier this month swore his oath of allegiance in Urdu, wearing traditional Pakistani clothes supplemented with a band of tartan. Elsewhere, less "visible" minorities, such as the Italians and now eastern Europeans, have stitched themselves into the fabric of Scotland's major cities, as indeed have Chinese groups and other east Asians. Hence one of Edinburgh's most beloved Scottish folk musicians is the Kirkcaldy-born Andy Chung.

Yet the most fascinating feature of Scottish nationalism is also the least noticed: there's a mighty difference between a nation's identity and people's national identities, which reveals itself in the saying that while England owned the British empire, it was the Scots who ran it. No wonder then that the Indian military has a Scottish tartan in its formal regalia (3rd battalion, the Sikh Regiment, traces its lineage from "Rattray's Sikhs" named after Captain Thomas Rattray of the 64th Regiment of Bengal Infantry). Whatever else people in Scotland think makes up their idea of Scottishness, the identity of Scotland as a historical nation cannot really be understood apart from that of India and other places of empire.

8. Why does the author use the example of Sikh shop owners selling Scottish paraphernalia?
A. To help explain the idea the concept of 'unlikely proponents of Scottish nationalism'
B. To raise the question of where ethnic and racial minorities fit into a historically 'white' country
C. To emphasise the concept that Scottish nationalism is not only for white Christians
D. To suggest that white Scottish people are intolerant of other ethnicities purporting to be Scottish nationalists
E. To ridicule the idea of multi-ethnic nationalism

9. What does the final paragraph of the extract conclude?
A. That there is a difference between people's individual national identities and the identities of a nation.
B. That a nation's identity and people within that country's national identity are entirely different concepts
C. The identity of Scotland is identical to those of other nations all around the world
D. That a nation's identity will have an effect on the national identities of individuals
E. That if Scotland loses the England as 'the other' it may turn inwards on its own minorities

10. All of the following are implied or stated in the piece except:
A. Scotland is becoming more multicultural
B. Scottish people are generally wary of their ethnic minorities
C. There has been very little writing done on ethnic minority Scots
D. Britain is established as a multicultural country
E. 'Englishness' as a concept is much less clear as a nationality than 'Britishness'

### 4. What price for honesty?

Adapted from 2014© Neil Reaich for BizEd

Pricing seems to be mentioned quite a lot in recent news headlines. The latest is the investigation by the Office of Fair Trading (OFT) into just how genuine the advertised price cuts in some large furniture stores really are. The OFT use the term 'reference pricing'. They've found cases in shops under investigation where not a single product had, in reality, been sold at the (supposedly original) higher price. The argument goes that since 95% of sales were at the lower or 'now' price then the stated original prices cannot really be genuine. Are we really so stupid as to be misled by reference pricing? When my wife buys some clothes in a sale and I ask how much it cost, she always responds by saying how much money she has saved. Well, unless she had the definite intention of buying it in the first place, she hasn't saved anything: in fact, she has spent it. The question is whether or not the reference pricing made a difference to her decision to buy the product. It must do, otherwise why is it so commonly used?

Heuristics suggests that people are not as rational as the standard economic model implies. Instead they use rules of the thumb, educated guesses or short cuts in decision-making. Anchoring refers to people making decisions based upon something they know to start with. For example, the anchor price of a jumper was £40 and there is 20 per cent discount offer making the new price £32 a saving of £8. If I decide to buy I might be thinking of the £8 saved because I didn't have to pay the full price, when I should be thinking logically about the actual price I am paying. In this case, there is too much focus on the anchor price, which is at such a level that I wouldn't have purchased the item anyway. In some cases, reference pricing is part of a strategy of price discrimination over time. Clothes shops attempt to capture consumer surplus by charging a high initial price for a few weeks for the 'must have and will pay' customers. The shops then give a discount which increases over time. New potential customers must now weigh up whether to buy now and get a 25 per cent discount or wait longer for a 5 percent saving and risk losing the deal because they have run out of their size.

Another tactic is to use time-limited offers where notice is given that the offer ends soon: buy now or miss out. Double glazing sales were notorious for using this tactic, encouraging customers to 'sign up today to get an extra 20 percent off'. Supermarkets are always giving volume offers, such as three for the price of two or get the second purchase for half the price. The supermarkets know we will pay more for the first item than we would for a second or third one. This tactic must lead to food waste as we are tempted to buy some products that we cannot possibly use before the sell-by date. The OFT also look at baiting sales where only a very limited number of products are available at the most discounted price. This sounds a little like discounts given for advanced booking train tickets. The OFT always makes mention of portioned 'drip' pricing where price increments drip through the buying process. It's those little add-ons that all add-up. Booking certain airline tickets comes to mind here. Restaurants often make an offer of a free second main course meal after buying one main course at full price. Now add on the fact that we may have full priced deserts, order drinks at a high mark up and then add the tip. Get your calculator out and the deal doesn't seem so good.

All these examples of price framing seek to alter a consumer's perception of the value of the offer. But there is more: have you purchased a printer at a ridiculously low price only to be stung on purchasing printer ink cartridges? Some businesses operate at a loss or low profit margin on certain items, to entice customers to part with more money on higher-profit goods. They bundle items together.

Some advice then:

- Use a calculator when shopping
- When buying clothes, divide the total by 50 to give you the average cost of wearing something once a week for a year: a shirt costing £25 will work out at 50 pence for every day worn;
- Avoid impulse buying on larger-ticket items;
- Only purchase multi-buys with long sell-by dates;
- Ignore the original price: it's only the current price you need to know about;
- Avoid limited-time offers.

11. What is the purpose of the inclusion of the phrase 'Booking *certain airlines* comes to mind here.'?
A. The author wants to leave the name of the airline open to individual inference, to make the statement as relevant as possible to all readers
B. The author wants to create a commonality between himself/herself and the reader, thereby making themselves more legitimate
C. The author wants to refer to a specific budget airline or group of airlines, but is fearful of naming one in particular
D. None of the above
E. All of the above

12. Which of the following is *not* a way that the author suggests retailers attempt to persuade customers into spending money on their products?
A. Stating that the 'original' price of an item is higher that it has ever really been sold, thereby creating the illusion of saving money
B. Splitting up the real cost of a product into a number of small, necessary, purchases
C. Raising the price of product - marketing as a 'luxury product' so that it will become more desirable
D. Giving some items for a reduced price if you buy in bulk
E. Slowly decreasing the price over time, creating a highly-pressured decision between buying now, or gambling that a product will be available at a cheaper price later

13. Why is it suggested that people are not as rational as the standard economic model implies?
A. To further the author's main point that people's decisions are influenced by different pricing models implemented by retailers
B. People make mistakes in their decision making - they aren't as thorough as they could be in the analytical process
C. To emphasise that the standard model is wrong, and that we must look beyond it to explain people's commercial behaviour
D. People are irrational, incapable of making decisions based on evidence
E. People are blind to price-based marketing

14. What is the overall purpose of the extract?
A. To advise consumers on how to make more rational decisions when shopping
B. To persuade retailers to be more honest in their pricing strategies
C. To warn consumers of the dangers of shopping without thinking about marketing
D. To inform readers about the different strategies utilised by retailers
E. To explore the concept of 'honest pricing' and come to a conclusion about the meaning of that term

### 5. A-level students: if you don't get into a Russell Group university, skip going altogether

Adapted from 2014© Christopher Giles for the Telegraph

*There is little point getting into thousands of pounds of debt for a degree in 'Make-up and Hair Design', writes Christopher Giles.*

Students who have received your A-level results, please listen up. If you have not – or do not – get into a top UK university, don't go at all. Results day is stressful. Euphoria. Relief. Disappointment. Frustration. Anxiety. It's emotional, but you'll be fine. Just follow my advice: if you aren't going to a Russell Group university or otherwise respected institution, forget about it altogether. Life is about making the right investments and university is pretty big one. It has to be worth every penny.

If, after the emotional roller coaster, you do decide to go to university, you are going to have a lot of debt. It is likely you'll owe £43,000, thanks to recent policies from the suits in Westminster. This will be money taken from your income for years to come (perhaps even for 30 years).

It is not just a financial investment. It is an investment of your time. If you live to 81, the average UK life expectancy, you would have spent 2.4 percent of your life at university. If this is 2.4 per cent of your life studying adventure sports, tourism or homeopathy, it probably won't propel your career. It is true that many have a love for academia – but if that's not you, don't be a sheep. Getting into and going to university used to be a big deal; now it's considered the norm. Don't let that fear of missing out sway you.

Time for some optimism: choosing not to go to university because you haven't got into one of the elite institutions shows good judgment. The appetite for mass university attendance is flagging. In fact, the stigma attached to having a weak degree could be worse than not having one at all. It is a nice sentiment to want everyone to go to university, but in reality it's not such a good idea.

To help you out, I've found some of the worst universities for graduate employability (via The Complete University Guide) and some strange courses to avoid. These universities really are the bottom of the pile and here is a selection of courses they offer, along with annual tuition fees.

The University of Sunderland: Fashion Product and Promotion (£7,800), Glass and Ceramics (£8,500), Sports Journalism (£8,500).

Southampton Solent University: Make-up and Hair Design, Football Studies, Cruise Industry Management. All of these courses cost £8,050.

The University of East London: Songwriting, Computer Game Design (Story Development), Hospitality and International Tourism Management. The University of East London ranks bottom in terms of graduate prospects in the UK and they charge £9,000 a year for full-time undergraduate courses.

Despite the huge debt, and wasted hours of your life, university does provide you with general life skills. Like the ability to work with a hangover, pay rent, and call your mum every weekend. Virtues these may be, but I am certain life experience isn't exclusive to university. It is just a costly way of getting it. Especially if what you're paying for is a tumbleweed degree that's going to float straight past employers.

The UK has some of the world's best universities. It also has quite a few average ones. So when you finally get through UCAS's 10,000-mile internet traffic jam, try to think with a cold rational mind: is it worth it?

15. What assumption does the author make in his discussion of university as 'an investment'?
A. Going to university will cost a significant amount of money
B. People will not be supported by family members in their financial commitment to pay university fees
C. Time spent at university is time wasted
D. Politicians do not have students' best interests at heart
E. There is no other reason why someone would choose to go to university except as an investment for the future

16. Which of the following is true according to the author?
A. University teaches people more than only the subject they study
B. A decision not to go to university is indicative of shrewdness
C. Since the rise in fees, university has become too expensive
D. A degree in 'Glass and Ceramics' is worthless
E. Only Oxbridge universities are worth attending

17. Which of the following is a possible description of the writer's opinion of the nature of the university system inferred from this passage?
A. Extortionate
B. Beneficial
C. Flawed
D. Worthless
E. Good value for money

### 6. Maria Miller: thank you and goodbye

Adapted from 2014© Simon Tait for 'The Stage' (edited)

She "has done an effective job in making the case for the value of public funding" Bazalgette says in response to the news that the arts are to get a ring-fenced 5% cut when ACE had been told to model for 10% and 15% scenarios for 2015/16.

But it is rather damning with faint praise. He goes on to say: "It is hugely encouraging that the Chancellor and the Treasury have listened to the argument that the arts and culture makes such a valuable contribution to our quality of life and the economy".

The argument made by the secretary of arts and culture, note, not the secretary of state.

DCMS (Department for Culture, Media and Sport) as a whole has got a less generous cut of 8%, and the deal for the arts appears to have been negotiated separately by the likes of Bazalgette and national museum directors like Nicholas Serota who, three weeks ago, went to George Osborne directly with economic arguments for a more lenient treatment of the sector. It was at this point that Osborne and the Treasury finally "got it" and realised how damaging a bigger arts cut would be to the economy, for negligible saving.

It means that Mrs. Miller cannot simply pass on to the arts the 8% cut as she and her predecessor, Jeremy Hunt, have done in the past because there is no fat in the DCMS, having been cut to the bone already, to absorb a new reduction itself. As it is, she will have to find the saving from elsewhere in her budget. Nevertheless, she has hung out and got a better settlement than most other government departments who are suffering at least 10% reductions as the government tries to find more savings, but it seems the knives are out not for culture or the arts but for Miller herself.

The knives appear to be out for her in government for a number of reasons, including not dealing decisively with Leveson. The culture secretary has also been the subject of unprecedented vilification in the Tory press, with the Daily Mail's drama critic Quentin Letts declaring a couple of weeks ago that "Culture is the department where a country can assert its character. If only its Secretary of State had one". In May, she made her first speech on the arts, calling for the economic argument to be made, Letts conceded, but "Where was the question of morality in Mrs. Miller's approach to the arts? Where was the vision that the arts can civilise us? Where was the idea of the arts as the most meritocratic of gifts, a route which can offer talented and aspiring youngsters a route to self-fulfillment…? There is not even much impression she is an arts lover. It was a speech that could have been given by any one of her departmental officials".

Her desperate attempts to grab a positive headline culminated last week in a damp squib of an announcement about the First World War centenary commemoration, in which nothing new was announced (except that 600-odd streets in England were to be renamed after VC winners from the Great War), and the major news about the cultural element cannot be revealed before August. On Friday, The Times's normally gentle columnist Richard Morrison wrote that "Some (culture secretaries) have been bores; some bluffers. But not one has depressed me as Maria Miller does.

As for the arts, the triumph is substantial and this might be a seachange in the way governments see the sector. The Arts Council, as fuel for the Bonfire of the Quangos, has taken an enormous battering since 2010 and the sector has correctly acknowledged that there is no case for "special treatment" while cuts amounting to 33% have been meted out, and of 50% to ACE itself. But now culture has established the principle that it is a special case after all, and with sense and imagination much of the effect of the new 5% cut might be ameliorated through the National Lottery.

The question now is whether that principle will be accepted by the other great subsidisers of the arts, the local authorities in whose hands the futures of dozens of theatres lie and whose extreme economic pain is even greater than Osborne's.

18. Which of these premises, if true, would most weaken the author's argument for the importance of funding for the arts?

A. Funding for the Emergency services is in decline and has been for years
B. The EU has recently passed legislation mandating that each country must increase state funding for the arts yearly in line with inflation
C. There is no evidence that the cost involved in increasing funding for the arts is justified by the eventual benefits
D. There is little evidence to suggest that funding for the arts would benefit a broad range of people
E. There is a correlation between increased funding for the arts and drop in the average GCSE grade

19. None of the following show the author's distaste for Maria Miller except:

A. 'not one has depressed me as Maria Miller does'
B. 'her desperate attempts to grab a positive headline'
C. 'it is rather damning with faint praise'
D. 'she has hung out and got a better settlement'
E. 'In May she made her first speech on the arts'

20. Which of the following is not stated in the extract?

A. The culture secretary now cannot pass cuts in budget straight onto the arts
B. Ms Miller is not universally accepted within government
C. The arts have had a better than expected outcome in relation to forecasts
D. The National Lottery's input will make up the whole of the 5% cut in the arts
E. Ms Miller announced a 'First World War centenary commemoration'

21. What, according to the author, is 'the question'?

A. Whether the idea that arts are more worthy than they once were is yet widely accepted
B. Whether the idea that arts are more worthy than they once were can be widely accepted
C. Whether the idea that arts are a 'special case' is correct given the many differing opinions
D. Whether the arts should be considered a 'special case'
E. Whether the idea that the arts should be treated differently will be reflected at a regional level

## 7. What makes a terrorist?

Adapted from 2014© Simon Allison for the Institute of Security Studies (edited)

Given the extent to which it dominates global news and politics, it is remarkable how little we know about the men – and, very occasionally, women – behind Islamic extremism. How are people drawn into such radical politics? What type of person becomes a terrorist? What is it that forces radicals out of day-to-day politics and into the extreme and often violent margins of society?

To discourage radicalisation we need a proper answer to the question of why and how people are radicalised. Anneli Botha, a senior researcher on terrorism at the Institute for Security Studies, asked members of radical groups themselves, specifically al-Shabaab and the Mombasa Republican Council.

Al-Shabaab is an Islamic extremist group that seeks principally to create an Islamic state in Somalia, although its ambitions and operations don't stop at the Somali border. The MRC, although often associated with al-Shabaab, is a distinct organisation with a very different agenda. It advocates for the secession of Kenya's coastal areas, and emphasises land grievances and economic and political marginalisation. It is a predominantly, but not exclusively, Muslim organisation.

95 interviews were conducted with individuals associated with al-Shabaab and 45 individuals associated with the MRC. Most of the respondents (96%) were male. From these interviews, Botha was able to observe patterns around their education, family background, religion and ethnicity as well as interrogate their motivations for joining and staying with radical groups.

There is no such thing as a typical extremist, however, it is possible to observe certain trends from the data. Almost all respondents grew up in a male-dominated household. Even when the father had passed away, usually a male relative stepped in to assume the patriarchal role. Over 70% of all respondents experienced corporal punishment at home, although most said that this wasn't too severe. Unexpectedly, around 60% of all respondents were middle children. "Middle children are known to experience the greatest sense of 'not belonging'," said Botha, explaining that this makes them especially vulnerable to close-knit radical groups, which can fill this void.

"When asked to clarify [what] finally pushed them over the edge, the majority of both al-Shabaab and MRC respondents referred to injustices at the hands of Kenyan security forces, specifically referring to 'collective punishment'," said Botha.

Respondents complained that "all Muslims are treated as terrorists" and that "government and security forces hate Islam". Some pointed to more specific examples, such as the assassination of Muslim clerics, or even particular incidents, such as an alleged assault by Kenyan police on a group of Muslims.

It is this last point that is most relevant for policymakers who are looking to contain the threat of extremism in Kenya. Simply put, a counter-terrorism strategy that relies on mass arrests, racial profiling and extrajudicial killings is counter-productive. These tactics have radicalised dozens, if not hundreds, of individuals, argues Botha, "ensuring a new wave of radicalism and collective resolve among their members, ultimately indicating that threats of violence or imprisonment are rarely effective deterrents."

Botha's research gives unprecedented insight into the background and motivations of the people who make up extremist organisations in Kenya. This is what a Kenyan radical looks like. Now it is up to Kenya's policymakers to tailor their responses accordingly.

22. According to the article, which of the following statements are accurate?
A. Extremists tend to share similar childhood experiences
B. Extremists are mostly male
C. Recent counter-terrorism strategies have led to increased radicalisation
D. All of the above
E. Both B and C

23. Based on the article, which of these pairs of ideas are not shown to be causally related?
A. Global coverage of radicalism and rise in extremism
B. Mass arrests and radicalism
C. Patriarchy and membership of extremist groups
D. An understanding of radicalisation and a reduction in radicalisation
E. Generalisation and frustration

24. What can be inferred from Botha's research as the definitive reason why most people become radicalised?
A. Patriarchal upbringing
B. Being a 'middle child'
C. Being overtly religious
D. None of the above
E. Impossible to say

25. Which of the following can be said to be the main conclusion of the final two paragraphs?
A. Radicalism is not always a result of indoctrination, background and childhood circumstances play a part too
B. The knowledge gained from this study about extremism in Kenya can be used to inform governments throughout the world
C. Policymakers should realise that most extremists identify unjust government activity as the ultimate reason why they became radicalised
D. Policymakers should take into consideration that the threat of extremism grows with an aggressive counter-terrorism policy
E. Policymakers should realise that extremism is best tackled with greater understanding of the reasons behind fanaticism

**8. Drugs laws play 'cat and mouse' with creators of legal highs, says senior government adviser**

Adapted from 2014© David Barrett for the Telegraph

Drug control legislation is being forced to play "cat and mouse" with the creators of an ever-expanding range of new "legal highs", the government's senior drug adviser has warned.

Professor Les Iversen, chairman of the Advisory Council on the Misuse of Drugs (ACMD), said the category of synthetic drugs – formally known as "novel psychoactive substances" – was growing all the time and legislation was struggling to keep up. He said the council, which advises the Home Secretary on classifying legal and illegal substances, had already reviewed the controls on legal highs in 2009 and 2012 before the launch of its current enquiry earlier this year.

"It looks as though we might have to go on doing this," he said. "It's a cat and mouse game but the cat should not withdraw in defeat."

Chemists who create synthetic drugs, such as mephedrone, to mimic the effects of illegal drugs like cocaine or heroin have been known to tweak the chemical composition of substances to stay one step ahead of the law. A European body which monitors new substances appearing on the black market said 80 different new types of legal high appeared last year.

Prof. Iversen added that the term "legal high" was now inappropriate because many of the substances, including mephedrone, have been made illegal. "The number of serious harms such as deaths emerging from this group of drugs is not all that staggering but it is an issue that we will certainly continue to consider," he said, adding that several announcements on the council's work in this area are due soon once they have secured ministerial approval.

The chairman also said during a meeting of the ACMD in central London that the growth of online pharmacies was leading to far wider abuse of prescription drugs which are legitimately available as painkillers or other types of medication.

"The misuse of medicines has become more of a topic because prescription medicines are so much more easily available on the internet," said the chairman, a retired Oxford University professor of pharmacology. "It's pretty easy to buy most medical products and this gives us a problem that is not easy to solve. We are going to try to evaluate the magnitude of the problem."

Prof. Ray Hill, chairman of the ACMD's technical committee, said it was currently looking at abuse of a drug called pregabalin which is used to alleviate pain and anxiety, and to treat epilepsy. "We are getting a variety of reports now from various parts of the UK that this drug is being misused," said Prof Hill.

"It's a drug that has some similar characteristics to opioid drugs when taken in very high doses. It's probably too soon to have substantive data on its harmful effects."
Prof. Iversen said: "The misuse of pregabalin is a surprise to all of us. This is a relatively new phenomenon and we're looking at how serious a problem this is."

26. Why is it suggested that the misuse of medicines is becoming more prevalent?
A. Growth of online pharmacies
B. Widespread use of the internet
C. The range of 'legal highs' is 'ever-expanding'
D. Legislation cannot keep up with chemists creating synthetic drugs
E. It is becoming easier to buy prescription drugs

27. Which of the following can be seen as being implied by Professor Iversen?
A. Legislators are giving up on the fight against legal highs
B. There is a labelling issue with so-called 'legal highs'
C. Pregabalin is being used more widely than before
D. The problem is becoming more difficult to deal with
E. The misuse of medicine is a longstanding issue

28. How many individuals are referred to in the extract?
A. 1
B. 2
C. 3
D. 4
E. 5

**9. The world needs to talk about child euthanasia: Mercy for all?**

Adapted from 2014© David Barrett for the New Scientist

EUTHANISING an infant is not technically difficult. Intravenous sedatives are used to silence the brain, followed by a pain medication such as morphine. This is often enough to trigger respiratory arrest and death, but if not, neuromuscular blockers are added, and the child dies. The process takes 5 to 10 minutes. Belgium has just become the first country to legislate in favour of child euthanasia at any age. However, there is a partial precedent. In 2005, the Netherlands recognised the Groningen protocol, a set of criteria outlining the circumstances under which ending the life of an infant under the age of 1 is permissible. Under those guidelines – which were written by Verhagen – euthanasia can only be undertaken if an infant's diagnosis and prognosis are certain and confirmed by an independent doctor, there is evidence of hopeless and unbearable suffering, both parents give their consent, the procedure follows medical standards, and all details are documented. Dutch children aged between 1 and 12 cannot be euthanised under any circumstances, although Verhagen and others are working to change that. While euthanasia remains technically illegal for infants in the Netherlands, doctors are not prosecuted so long as the protocol's criteria are met.

Opponents argued that this would lead to a slippery slope of infant euthanasia. The opposite happened. Since 2005, there have been only two cases in the Netherlands. Both involved babies with lethal epidermolysis bullosa, a disease of the connective tissues. This decline in euthanasia correlates with an increase in late-term abortions. Previously, most euthanasia cases involved babies born with severe to extreme spina bifida – a congenital disorder in which some of the vertebrae do not fully form. Doctors found that surgery was not possible and that the child would suffer constantly. In 2007, the Netherlands began offering free ultrasound scans at 20 weeks of pregnancy, at which point spina bifida can be detected. Mothers whose babies are diagnosed with the disease can then decide whether to terminate the pregnancy. This is not necessarily the best course for everyone in this situation. Only the most extreme cases of spina bifida are deemed hopeless, and it is impossible for doctors to precisely gauge the severity in utero. Having infant euthanasia as an option allows mothers to be sure that their baby has no chance of survival before ending its life. But, as Verhagen says, in practice most in this situation decide not to take any chances and terminate the pregnancy.

The means of ending a baby's life are subject to debate. Recently, the line between proactive palliative care – applying pain medications that may hasten death – and euthanasia has become more blurred. In some countries, including the US, food and fluid may be withdrawn in some circumstances. But palliative care practices do not necessarily result in a quick death for a terminally ill infant. Death by dehydration and starvation can take days or weeks and it is impossible to guarantee that the child – even heavily medicated – does not suffer. Moreover, no one doubts that death is the outcome of withholding life-sustaining care and support. Rather than draw out the inevitable, would it not be less cruel to swiftly end that life, alleviating all risk of unnecessary suffering? Belgium and the Netherlands have chosen to face this dilemma directly. Of course, not every country is as progressive, there will always be those who – due to religious or personal beliefs – oppose ending a human life. In the US, for example, reaching a federal consensus on the subject of infant euthanasia seems unlikely. On the other hand, progressive states such as Oregon might someday implement their own laws on it, much as they have for assisted suicide in adults. Whether this will ever come to pass remains to be seen. As Verhagen wrote in *The New England Journal of Medicine*, "This approach suits our legal and social culture, but it is unclear to what extent it would be transferable to other countries." For most parts of the world, a refusal to even discuss the subject dominates. As unpleasant as it is, parents, physicians, hospitals and nations need to confront this issue as a matter of responsibility towards both infants born into hopeless circumstances and their families.

29. What assumption does the author make in the last paragraph?
A. That late term abortion has recently become legalised
B. That fathers do not have a say in whether or not a child is to be aborted
C. It is a positive step that infant euthanasia has decreased in frequency
D. Epidermolysis Bullosa will always lead to early death
E. Mothers who terminate a pregnancy without being sure that their foetus' disease will be fatal are negligent

30. What is the main conclusion of the article?
A. Spina bifida is a horrific disease
B. Infant euthanasia should be widely legalised
C. Late term abortion can pre-empt a logical decision on the continuation of an infant's life
D. More of an agreement should be reached about the benefits of legalisation of child euthanasia in certain, narrow, circumstances
E. Infant euthanasia should be discussed more openly

31. Which of the following (if true) strengthens the author's argument?

A. Spina bifida's survival rate for newborns is 10%
B. Upon hearing about the inhumane palliative care treatment, 70% of parents with children with terminal, painful illness would choose euthanasia above practice palliative care options
C. Belgium's child mortality rate has risen by 25% since the legalisation
D. Spina bifida has been proven to cause continuous excruciating pain to infants from the moment of birth
E. The President of the European Union has given a speech suggesting that an investigation should be conducted into the possible benefits and problems of extending palliative care practices to cover child euthanasia

32. What is most likely to be the reason that the author begins the article with the sentence 'Euthanizing a child is not technically difficult.'?
A. Stating that euthanasia is not *technically* difficult implies that the real problems lies in its *emotional* difficulty
B. To grab the readers' attention at the beginning of the extract
C. To inform the reader about the methods used in infant euthanasia
D. None of the above
E. All of the above

### 10. Brazil releases 'good' mosquitoes to fight dengue fever

Adapted from 2014© Julia Carneiro for the BBC

The hope is they will breed, multiply and become the majority of mosquitoes, thus reducing cases of the disease.

The initiative is part of a programme also taking place in Australia, Vietnam and Indonesia.

The intracellular bacteria, Wolbachia, being introduced cannot be transmitted to humans.

The programme started in 2012 says Luciano Moreira of the Brazilian research institute Fiocruz, who is leading the project in Brazil. "Our teams performed weekly visits to the four neighbourhoods in Rio being targeted. Mosquitoes were analysed after collection in special traps. "Transparency and proper information for the households is a priority."

Ten thousand mosquitoes will be released each month for four months with the first release in Tubiacanga, in the north of Rio.

**'Good' bacteria**

The bacterium Wolbachia is found in 60% of insects. It acts like a vaccine for the mosquito which carries dengue, Aedes aegypti, stopping the dengue virus multiplying in its body.

Wolbachia also has an effect on reproduction. If a contaminated male fertilises the eggs of a female without the bacteria, these eggs do not turn into larvae. If the male and female are contaminated or if only the female has the bacteria, all future generations of mosquito will carry Wolbachia.

As a result, Aedes mosquitoes with Wolbachia become predominant without researchers having to constantly release more contaminated insects.

In Australia this happened within 10 weeks on average. The research on Wolbachia began at the University of Monash in Australia in 2008. The researchers allowed the mosquitoes to feed on their own arms for five years because of concerns at the time Wolbachia could infect humans and domestic animals.

Three more neighbourhoods will be targeted next, and large scale studies to evaluate the effect of the strategy are planned for 2016.

Dengue re-emerged in Brazil in 1981 after an absence of more than 20 years. Over the next 30 years, seven million cases were reported. Brazil leads the world in the number of dengue cases, with 3.2 million cases and 800 deaths reported in the 2009-14 period.

33. What is the ultimate aim of the project?
A. To release 'good' mosquitoes
B. To reduce the number of people suffering from Dengue Fever
C. To reduce the number of mosquitoes carrying Dengue Fever
D. To reduce the number of people suffering from diseases carried by mosquitoes
E. To follow in Australia's footsteps

34. Which of the following statements would be most damning to the decision to release 'good' mosquitoes?
A. The discovery that Wolbachia has the capacity to mutate such that it can be transmitted to, and will be fatal to, humans
B. The discovery that Wolbachia does not act as a vaccine for Aedes aegypti
C. The discovery that Wolbachia can only be transmitted from male to male mosquitoes
D. The discovery that the evidence from Australia that showed that Wolbachia could not infect humans was tampered with
E. The discovery that Wolbachia can be transmitted to, and causes infertility in, humans

35. Which of the following is not addressed as a potential (but surmountable) issue when releasing mosquitoes?
A. Members of the public panicking about the sudden influx of mosquitoes
B. Wolbachia being transmitted to humans
C. The whole process taking longer than 10 weeks
D. The need for more and more mosquitoes to be released periodically
E. Wolbachia being transmitted to livestock

## 11. If you're a feminist you'll be called a man-hater. You don't need rebranding

Adapted from 2014© Laurie Penny for the Guardian (edited)

Nobody likes a feminist. At least not according to researchers at the University of Toronto, following a study where it emerged that people still defer to stereotypes about "typical" feminist activists, stereotypes including "man-hating" and "unhygienic". These stereotypes are apparently seriously limiting the appeal of women's liberation as a lifestyle choice. Feminism is a mess, and needs to sort itself out. In order to be "relevant to young women today" it needs to shave its legs and get a haircut.

Elle seems to think so too. The fashion and beauty magazine, not a historically notable manual for gender revolution has weighed in this month with a spread on "rebranding feminism", asking three advertising agencies to give gender politics a nip, tuck and polish. The result is flowcharts and a lot of hot pink equivocation that airbrushes out the ugly, uncomfortable bits of women's liberation. They'd prefer us to consider men's feelings first when we speak about work, pay and sexual violence, to be less threatening, to dress it up; they'd prefer us to talk about "equalism" if we must speak at all. Those with a vested interest in the status quo would prefer young women to act more like they're supposed to – to make everything, including our politics, as pretty and pleasing as possible.

The rebranding of feminism as an aspirational lifestyle choice, a desirable accessory, as easy to adjust to as a detox diet and just as unthreatening, is not a new idea. Nor is ELLE magazine even the first glossy to attempt the task in recent years. But unfortunately there's only so much you can "rebrand" feminism without losing its essential energy, which is difficult, challenging, and full of righteous anger. You can smooth it out and sex it up, but ultimately the reason many people find the word feminism frightening is that it is a fearful thing for anyone invested in male privilege. Feminism asks men to embrace a world where they do not get extra special treats merely because they were born male. Any number of jazzy fonts won't make that easy to swallow.

It is not "young women today" who need to be convinced that feminism remains necessary and "relevant". Changing technology has shaken up a tsunami of activism around gender and politics, from initiatives like the Unslut project and Everyday Sexism to sea changes in culture like the backlash against sexual violence in India. In all of these movements, young women are leading the charge, along with a few fighters from older generations who have not been worn down by decades of mockery and marginalisation. While the fashion press and the beauty industry remain invested in the idea of young women as pliant, affable and terminally anxious about getting boys to like them, real women and girls are fighting back against a culture that persists in trying to present our desires and rebrand our politics as fluffy and marketable.

The stereotype of the ugly, unfuckable feminist exists for a reason – because it's still the last, best line of defence against any woman who is a little too loud, a little too political. Just tell her that if she goes on as she is, nobody will love her. Correct me if I'm wrong, but I've always believed that part of the point of feminist politics – part of the point of any sort of radical politics – is that some principles are more important than being universally adored, particularly by the sort of men who would prefer women to smile quietly and grow our hair out.

In the words of the early suffragist and civil rights campaigner Susan B Anthony: "Cautious, careful people always casting about to preserve their reputation or social standards never can bring about reform. Those who are really in earnest are willing to avow their sympathies with despised ideas and their advocates and bear the consequences."

I am not so very old, but I'm old enough to have noticed that the times in my life when I was most admired by men, the times when I was considered most likeable, were also the times when I was most vulnerable, most powerless and unsure of myself. The times when I've been strongest and most daring, the times when I've been proudest of my own achievements – that's when I've been called a difficult bitch. That's what women get to choose, now as much as at any point in history: how much we are willing to sacrifice to make men like us.

36. All of the following words and phrases suggest disapproval except:
A. 'seriously limiting the appeal of women's liberation'
B. 'has weighed in this week'
C. 'how much we are willing to sacrifice to make men like us'
D. 'the sort of men who would prefer women to smile quietly'
E. 'a lot of hot pink equivocation that airbrushes out the ugly'

37. Why might Laurie Penny uses the word 'jazzy' when describing fonts?
A. To explain to the reader what the fonts look like
B. To draw attention to the illogicality of making feminism more digestible by using the juxtaposition of a seemingly uncomplicated and simple word 'jazzy' against the inherently complex idea of feminism
C. To draw attention to the idea that 'sexing' up the idea of feminism is laughable
D. To belittle what fashion magazines are trying to do
E. All of the above

38. Which of the following phrases can be said to be sarcastic?
A. 'hot pink equivocation that airbrushes out the ugly, uncomfortable bits of women's liberation'
B. '"typical" feminist activitists'
C. 'not a historically notable manual for gender revolution'
D. None of the above
E. All of the above

39. Which of the following is not a description of a feminist found in the extract?
A. Difficult
B. Unsanitary
C. Pliant
D. Relevant
E. Unattractive

### 12. Complex organic molecule found in interstellar space

Adapted from 2014© Michael Eyre for the BBC (edited)

*Scientists have found the beginnings of life-bearing chemistry at the centre of the galaxy.*

Iso-propyl cyanide has been detected in a star-forming cloud 27,000 light-years from Earth. Its branched carbon structure is closer to the complex organic molecules of life than any previous finding from interstellar space. The discovery suggests the building blocks of life may be widespread throughout our galaxy.

Various organic molecules have previously been discovered in interstellar space, but i-propyl cyanide is the first with a branched carbon backbone; this is important as it shows that interstellar space could be the origin of more complex branched molecules, such as amino acids, that are necessary for life on Earth.

Dr Arnaud Belloche from the Max Planck Institute for Radio Astronomy is lead author of the research, which appears in the journal Science. "Amino acids on Earth are the building blocks of proteins, and proteins are very important for life as we know it. The question in the background is: is there life somewhere else in the galaxy?"

**Watch the skies**

The molecule was detected in a giant gas cloud called Sagittarius B2, an active region of ongoing star formation in the centre of the Milky Way. As stars are born in the cloud they heat up microscopic dust grains. Chemical reactions on the surface of the dust allow complex molecules like i-propyl cyanide to form. These molecules emit radiation that was detected as radio waves by twenty 12m telescopes at the Atacama Large Millimeter Array (Alma) in Chile.

Each molecule produces a different "spectral fingerprint" of frequencies. "The game consists in matching these frequencies... to molecules that have been characterised in the laboratory," explained Dr Belloche. "Our goal is to search for new complex organic molecules in the interstellar medium, the ultimate aim is to know whether the elements that are necessary for life to occur... can be found in other places in our galaxy."

Prof. Matt Griffin, head of the school of physics and astronomy at Cardiff University, commented on the discovery. "It's clearly very high-quality data - a very emphatic detection with multiple spectral signatures all seen together.

"There seems to be quite a lot of it, which would indicate that this more complex organic structure is possibly very common, maybe even the norm, when it comes to simple organic molecules in space."

"It's a step closer to discovering molecules that can be regarded as the building blocks or the precursors... of amino acids." The hope is that amino acids will eventually be detected outside our Solar System. "That's what everyone would like to see," said Prof Griffin. If amino acids are widespread throughout the galaxy, life may be also.

"So far we do not have the sensitivity to detect the signals from [amino acids]... in the interstellar medium," explained Dr Belloche. "The interstellar chemistry seems to be able to form these amino acids but at the moment we lack the evidence. [...] Alma in the future may be able to do that, once the full capabilities are available."

Prof Griffin agreed this could be the first of many further discoveries from the "fantastically sensitive and powerful" Alma facility.

40. Which of the following best describes Professor Griffin's opinion on the implication of the discovery?
A. It is unremarkable
B. It is a reliable conclusion
C. It is a significant discovery
D. It is unexpected
E. It could be a catalyst for further discovery

41. What is Iso-propyl cyanide?
A. Warmed microscopic dust grains
B. A molecule with a branched carbon backbone
C. Radioactive atoms
D. An organic molecule carrying life
E. A component of an amino acid

42. What can be inferred about amino acids from the extract?
A. In the future, amino acids may be identified outside of the Milky Way
B. The Alma facility could make discoveries at the forefront for molecular science
C. Amino acids are radioactive
D. Amino acids give off signals which are currently undetectable
E. Amino acids are essential for discovering space

**END OF SECTION**

**YOU MUST ANSWER ONLY ONE OF THE FOLLOWING QUESTIONS**

Your answer should be a well reasoned argument, showing your understanding of the issue in question and giving the reasons for any opinions you hold.

1. "The government should legalise the sale of human organs" Discuss

2. "Developed countries have a higher obligation to combat climate change than developing countries" Discuss the extent to which you agree with this statement

3. '"Putin is a serious threat to global stability" Discuss

4. "Sufferers of anorexia nervosa should be force fed" Do you agree with this statement? If so, evaluate at what point of an individual's disease this measure should be taken.

**END OF TEST**

THIS PAGE HAS BEEN

INTENTIONALLY LEFT BLANK

# Mock Paper B

### 1. Water Wars: Egyptians Condemn Ethiopia's Nile Dam Project

Adapted from 2013 © Peter Schwartzstein for National Geographic

As the Grand Ethiopian Renaissance Dam takes shape, tempers rise.

"Ethiopia is killing us," taxi driver Ahmed Hossam said, as he picked his way through Cairo's notoriously traffic-clogged streets. "If they build this dam, there will be no Nile. If there's no Nile, then there's no Egypt."

Projects on the scale of the $4.7 billion, 1.1 mile long (1.7 kilometre long) Grand Ethiopian Renaissance Dam often encounter impassioned resistance, but few inspire the kind of dread and fury with which most Egyptians regard plans to dam the Blue Nile river.
Egypt insists Ethiopia's hydroelectric scheme amounts to a violation of its historic rights, a breach of the 1959 colonial-era agreement that allocated almost three-fourths of the Nile waters to Egypt, and an existential threat to a country largely devoid of alternative freshwater sources.

But what Egyptians regard as a nefarious plot by its historic adversary to control its water supply, Ethiopians see as an intense source of national pride and a symbol of their country's renewal after the debilitating famines of the 1980s and '90s.

"People are enthusiastic. They're excited, because no leader has tried such a project in Ethiopia's history," said Bitania Tadesse, a recent university graduate from the capital, Addis Ababa. "It's a big deal that is going to be beneficial to future generations."

Ethiopia maintains that Egypt and Sudan downstream have no reason to be fearful. The government says it's merely redressing the inequalities of previous water-sharing arrangements, which had left the nine upstream countries largely bereft of access to the Nile.

1. What is implied by the Ethiopian government in the last paragraph?
A. Egypt and Sudan have treated Ethiopia badly in the past
B. Ethiopia has been overlooked in previous water sharing schemes
C. Ethiopia has been purposefully disadvantaged as a result of previous water sharing schemes
D. The Dam is but one retaliation in a long-standing feud between Ethiopia and Egypt
E. Egypt has less need for access to water than Ethiopia does

2. None of the following convey Egypt's distaste for Ethiopia except:
A. 'People are enthusiastic'
B. 'Egypt insists Ethiopia's hydroelectric scheme amounts to a violation of its historic rights'
C. 'Ethiopia is killing us'
D. '(I)mpassioned resistance'
E. 'Egyptians regard as a nefarious plot'

3. Why does Ahmed Hossam use exaggeration in the phrase 'If there is no Nile, there is no Egypt'?
A. He believes that Egypt will cease to exist as a State if the Dam is built
B. He uses stark terms to illustrate how strongly he feels about Ethiopia's decision to build the Dam
C. He hopes to explain how important the Nile is to Egyptian culture
D. He wants to emphasise the severity of the potential problem caused by the Dam
E. He uses sarcasm to belittle the fears of other Egyptians

## 2. Who is to judge which lives are worth living?

Adapted from 2011 © Barbara Ellen for The Guardian (edited)

*The able bodied should never dictate the fates of the ill and weak*

There has been some fuss about the forthcoming BBC documentary, *Choosing to Die*, presented by novelist and Alzheimer's sufferer Terry Pratchett, which features a man with motor neurone disease, travelling to Swiss clinic Dignitas and – a first on terrestrial television – dying on screen.

The BBC has been accused of acting "like a cheerleader for legalising assisted suicide", which it denies. Pratchett says: "Everybody possessed of a debilitating and incurable illness should be allowed to pick the hour of their death." Clearly, with him, the dignity of choice is paramount. However, while one has enormous sympathy for Pratchett suffering such a vile disease, the fact remains that he is a rich, powerful man and it is highly unlikely that his wishes would be ignored. With respect, euthanasia laws are not in place to protect people such as him. What of those who may have their "choice" taken away, even if they don't want to die?

There are bigger issues at stake, not least the arrogance of the pro-euthanasia able bodied towards the profoundly ill – the unseemly rush to pronounce the lives of others "not worth living". A recent study discovered that some sufferers of locked-in syndrome – as many as three out of four of the main sample – were happy and did not want to die. Such studies are flawed (some sufferers are unable to articulate either way), but it should still give us pause for thought before blasting off about "lives not worth living".

Likewise the knee-jerk: "They wouldn't have wanted to end up like this." Of course not – who would? – but that might not be the end of the story. How individuals feel when they are fit may change considerably when their health fails. Like those with locked-in syndrome, they may adjust to a life that is very different, often difficult, but just as precious. Who are we to judge?

Bizarrely, the one thing the pro- and anti-euthanasia lobbies have in common is an obsession with God. Sometimes, it's almost as if the antis are tricked into talking about the "sanctity of life" and "God's will", to make the pros look more modern and credible.

Personally, if I ever get something nasty, I'd rather be with a God-botherer than somebody who decides I'm looking peaky, books a Swiss flight and whisks me off to the ghouls at Dignitas. Or maybe I wouldn't – maybe I'd be begging for death. The hope is that I'll choose.

At the moment, assisted death is illegal in Britain, with the caveat that each case is assessed individually, with empathy for the individual and their carers. It could be worse. One reason we don't have the death penalty is that there is no guarantee that mistakes would not be made. Who could guarantee that mistakes wouldn't be made with euthanasia? Not all seriously ill people can communicate their current wishes (not necessarily the same as when they first became ill). And no one else should be deciding for them, in worst-case scenarios "putting them down" against their will.

The phrase: "It's what they would have wanted" belongs after death, not before it. A prolonged, pointlessly agonising end is everyone's nightmare, but that doesn't mean the able bodied should ever get to dictate the fates of the ill and weak. Terry Pratchett should be commended for speaking up for those who wish to die with dignity. However, others who might not want to die, but can't articulate that, need a voice to speak up for them too.
Is this man dead or just dead stupid?

4. What is the main conclusion of the article?

A. The conclusion of Terry Pratchett's documentary is skewed by his own personal circumstances
B. Euthanasia is not yet a precise science and legalising it may give rise to irrevocable mistakes being made
C. Those who have the means to be euthanised should be able to do so if they wish to.
D. Euthanasia should be a personal choice, and legalising euthanasia may give rise to a situation where the seriously ill are robbed of their choice by the healthy
E. Euthanasia could only work if all seriously ill people could articulate their intention to die in such a way that it could not be misunderstood

5. What can be inferred from the discussion of Terry Pratchett's situation?

A. A less wealthy person's wishes are more likely to be ignored
B. Mr Pratchett does not care about those whose choice has been taken away
C. For Mr Pratchett, choice is primarily important
D. A wealthier person's wishes are less likely to be ignored
E. Mr Pratchett is not best placed to discuss any Euthanasia laws

6. Which of the following, if true, would rectify the main problem the author identifies in relation to euthanasia?

A. A law is passed requiring that all people must choose when healthy what they wish to happen to them if they become ill (at varying stages of illness)
B. A new medical procedure is discovered which can analyse the brain's electrical impulses and determine how content a seriously ill patient is
C. Euthanasia is made illegal worldwide, so that even the rich and powerful cannot choose to die
D. A study is published, concluding that 85% of people in a persistent vegetative state will never recover brain function
E. A scientific discovery means that people can be kept alive using life support machine for a fraction of the current cost

7. Why does the author use the phrase 'knee jerk'?

A. To emphasise the importance of evaluating an argument before one makes it
B. To cast doubt upon the value of the argument that follows it
C. To show that, when discussing euthanasia, often people have automatic reactions without truly analysing why they feel this way
D. To suggest that people's reflexive reactions to euthanasia are often wrong
E. To imply that there is no need to make the argument that follows due to it being universally acknowledged to be true

### 3. Illegal downloaders are one of music industry's biggest customers

Adapted from 2009 © Demos.co.uk

People who use peer-to-peer filesharing websites like Pirate Bay to illegally download music spend over £30 more on music per year than those who do not download illegally.

Internet users who claim to never illegally download music spend an average of £44 per person on music per year, while those who do admit to illegal downloading spend £77, amounting to an estimated £200m in revenue per year.

A new poll commissioned by Demos found that almost one in ten adults (9%) aged 16-50 who have internet access admit downloading music illegally. But this group are also active music buyers, with 8 in 10 buying CDs, vinyl or MP3s in the past year. The poll also found that 42 percent of illegal downloaders agree that they 'like to try things out before I decide whether to buy them.'

The findings suggest that government plans to disconnect repeat illegal downloaders from the internet, announced yesterday by Lord Mandelson, could do the music industry more harm than good by punishing core consumers. The poll showed that the availability of new, appealing legal music provision services is the step most likely to encourage illegal downloaders to stop, above fines or the threat of disconnection.

The research reveals a gap between what consumers are willing to pay for music tracks and current market prices. If official music distribution sites like iTunes lowered the cost of a single track to 45p they could expect prospective buyers to double in number.

Peter Bradwell, a researcher at Demos specialising in digital rights and consumer trends said:

"The latest approach from the government will not help to prop up an ailing music industry. Politicians and music companies need to recognise that the nature of music consumption has changed and consumers are demanding lower prices and easier access to music."

8. Which of the following is true, according to the article?

A. The Government's plans are destined to fail

B. Illegal downloading is not as big a problem as first imagined

C. The poll had surprising results

D. Most people who admit to illegally downloading music also buy music

E. All of the above

9. Which statement (if true) would be most damning to the validity of the poll?

A. Demos' CEO has been charged with tax evasion

B. The participants in the poll were friends of Peter Bradwell

C. Only 25 people were questioned

D. A typo in the questionnaire puts into question how many participants understood exactly what they were being asked

E. Half of the participants were bribed by the government's opposition to lie

10. Which of the following is an assertion of opinion rather than a statement of fact?

A. 'The poll showed that the availability of new, appealing legal music provisions is the step most likely to encourage illegal downloaders to stop

B. 'Politicians and music companies need to recognise that the nature of music consumption has changed'

C. 'The research reveals a gap between what consumers are willing to pay for music trackers and current market prices'

D. '8 in 10 buying CDs, vinyl or MP3s in the last year'

E. 'If official music distribution sites like iTunes lowered the cost of a single track to 45p they could expect prospective buyers to double in number'

### 4. Heavy Military Recruitment at High Schools Irks Some Parents

Adapted from 2005 © Kelly Beaucar Viahos for FoxNews.com (edited)

A little-known provision in the No Child Left Behind Act that compels public high schools to open their doors and pupil records to military recruiters has some parents, students and anti-war groups up in arms. "We think most people were unaware of it," Amy Hagopian, co-president of the Garfield High School Parent-Teacher-Student Association in Seattle and an active counter-recruiter in the school, said of the provision.

Hagopian said parents are just becoming aware of the policy, which gives recruiters the same access to high school campuses and students' phone numbers and addresses as colleges and businesses have. Districts that don't comply could risk annual federal funding. According to the law, parents must be notified and can refuse to release their children's information. Every school has adopted different notification policies, some being more effective than others, school officials said. Military recruitment issues have been making headlines in recent weeks, as the Army, Marine Corps and National Guard have announced shortfalls in their goals this year. Reports say recruitment pressure is translating into inappropriate tactics by recruiters to the extent that the Army halted recruiting for one day in May to refresh staff with proper protocol in dealing with prospective soldiers.

Paul Rieckoff, an Iraq war veteran and founder of Operation Truth, a veterans' advocacy organization, said parents are now reacting to "major recruiting problems" and bad news coming out of Iraq. "I think it's safe to say there is concern and even the beginning of a movement to combat or to face the recruiters at the high schools," he said. "We don't necessarily endorse that but the critical issue is that the Army has missed their goals again this year." The military has always had access to schools but not all have opened their doors and records equally. Now, the No Child Left Behind Act emboldens efforts to gain "access to the best and brightest this country has to offer," said Department of Defence spokeswoman Lt. Col. Ellen Krenke.

"For some of our students, this may be the best opportunity they have to get a college education," wrote Secretary of Defence Donald Rumsfeld and former Secretary of Education Rod Paige in an October 2002 letter to school superintendents announcing the new law. But some parents and teachers say school is not an appropriate place for the military's message, and complain the hard sell has gotten harder since the Iraq war began and following lacklustre recruitment numbers. "The recruiters really harangue people, and this is what parents are trying to avoid," said Tina Weishaus, president of the Highland Park Middle School/High School Parent Teacher Organization in New Jersey. "Personally, I think the whole thing should be struck from No Child Left Behind," Weishaus said. "I don't think the federal government should be mandating that schools become a recruiting ground for the war."

Army spokesman Doug Smith said the Army has not accelerated recruitment at the schools in the face of missed goals. It is primarily targeting college students, he said, with the average age of 21 for new Army recruits. In 2002, 12,560 out of the 77,000 enlistments were recruited out of high schools. Other critics say they have no problem with military recruiters, but are concerned about students' privacy. According to Montclair school officials, more than 80 percent of the parents who responded to that campaign asked that their records not be given to recruiters this year.

Bill Cala, superintendent of the Fairport Central School District in New York, said his school has been found non-compliant with the law because it doesn't release the names of students to the military unless parents specifically give their consent.

He said about 80 out of the 1,600 students in the school consented this school year, but recruitment among seniors hit 2 percent. "This really, for us, is a privacy issue and doesn't have anything to do with support for the military or for the war," Cala said.

11. Which of the following is a provision of the 'No Child Left Behind' Act?
A. That high schools must give over all the contact information of all their students to military recruiters
B. That high schools are given the option to hand over some students' contact information
C. That high schools must hand over all contact information unless the parents specifically withhold their consent
D. That high schools must give over all students' contact information provided that they have consent from the students to do so
E. That high schools must seek consent from parents before giving over students' contact information to recruiters

12. All of the following are opinions stated in the piece except:
A. The main concern about this provision centres around the students' privacy
B. School is not the right place for military recruitment
C. Recruitment tactics have become inappropriate
D. The recruiters' approach is too aggressive for children in a school environment
E. Most parents are unaware of this specific provision of 'No Child Left Behind'

13. What is the purpose of the inclusion of the phrase 'the hard sell has gotten harder'?
A. The author uses repetition to draw the reader's attention to the idea that recruitment tactics are increasingly rough
B. The author plays on the term 'hard sell' to add variation to the language of the piece
C. The author attempts to be facetious by making fun of the term 'hard sell' to belittle the parents' concerns
D. None of the above
E. All of the above

14. What can be inferred from Bill Cala's figures concerning the number of his students who consented to having their details shared with the military?
A. Most students do not want their details shared with the military
B. Seniors are more likely to consent to having their details shared than younger students
C. Current recruitment strategies are more likely to succeed with older students
D. All of the above
E. Both A and C

### 5. Found: giant spirals in space that could explain our existence

Adapted from 2015 © Michael Slezak for New Scientist (edited)

Giant magnetic spirals in the sky could explain why there is something rather than nothing in the universe, according to an analysis of data from NASA's Fermi space telescope.
Our best theories of physics imply we shouldn't be here. The Big Bang ought to have produced equal amounts of matter and antimatter particles, which would almost immediately annihilate each other, leaving nothing but light.

So the reality that we are here – and there seems to be very little antimatter around – is one of the biggest unsolved mysteries in physics.

*Monopole monopoly*

In 2001, Tanmay Vachaspati from Arizona State University offered a purely theoretical solution. Even if matter and antimatter were created in equal amounts, he suggested that as they annihilated each other, they would have briefly created monopoles and antimonopoles – hypothetical particles with just one magnetic pole, north or south.

As the monopoles and antimonopoles in turn annihilated each other, they would produce matter and antimatter. But because of a quirk in nature called CP violation, that process would be biased towards matter, leaving the matter-filled world we see today.
If that happened, Vachaspati showed that there should be a sign of it today: twisted magnetic fields permeating the universe – a fossil of the magnetic monopoles that briefly dominated. And he showed they should look like left-handed screws rather than right-handed screws.

So Vachaspati and his colleagues went looking for them in data from NASA's Fermi Gamma ray Space Telescope. As gamma rays shoot through the cosmos, they should be bent by any magnetic field they pass through, so if there are helical magnetic fields permeating the universe that should leave a visible mark on those gamma rays.

*All of a twist*

Lo and behold, that's just what they found – well, maybe. "What we found is consistent with them all being left-handed," says Vachaspati. "But we can't be sure." He says there's less than a one per cent chance that what they see in the Fermi data happened by chance. "That's being conservative," he says.

They also found that the twists in the field are a bit bigger than they predicted. "So there is some mystery there," says Vachaspati. He says more data from Fermi, which is expected this year, will help narrow down the odds.

Nicole Bell from the University of Melbourne in Australia warns that magnetic fields could have been caused in other ways, including from inflation. What's more, for CP-violation to provide enough matter in the universe you usually need "new physics" – stuff beyond the standard model of particle physics – which hasn't been confirmed experimentally yet. "But it is a very interesting idea," she says.

15. Which of the following best describes the author's stance in the first paragraph?
A. Human existence is inexplicable
B. Current theories suggest we should not be in existence
C. Our existence can be explained by an analysis of data from NASA's Fermi space telescope
D. The quantities of matter and antimatter particles created during the Big Bang can best explain our existence.
E. Physics will never be able to explain why humans exist

16. The presence of which of the following could indicate helical magnetic fields permeating the universe?
A. Antimatter particles
B. Magnetic monopoles
C. Gamma rays which have been bent by any magnetic field they pass through
D. A mark left on the magnetic field
E. All of the above

17. What does the author mean by "new physics"?
A. Theories beyond the normal model of particle physics
B. Theories which haven't been confirmed by experiments yet
C. Theories which are currently beyond our understanding
D. A and C
E. A and B

### 6. Making an opinion illegal is not going to stop terrorism

Adapted from 2015 © Jonathan Russell for The Telegraph (edited)

*If 'British values' mean anything, they must prevent us from legislating against non-violent views that we find abhorrent*

The Government's shift in focus to target extremism and tackle radicalisation rather than counter terrorism is a welcome move, but I question whether the measures set out by the Prime Minister will do this effectively, proportionately and in a way that befits our "British values". Among other things, the proposals are to include "extremism disruption orders", which would not criminalise the act of hate speech or promotion of terrorism, for they are already illegal under the Terrorism Act 2006, but rather the intent to do so, if ministers reasonably believe this to be the case. In essence, these measures target those who operate in what the police have called the "pre-criminal space" and therefore expand the definition of people who could be incarcerated from those who do bad things to those who think bad things. This is problematic for a number of reasons. First, we have the ethical issue of clamping down on freedom of expression, one of our universally accepted human rights. The risk is that the measures are likely to be used in instances when contemplation never graduated to, nor was ever going to graduate to, action. Our courts will have to prosecute entirely on *mens rea* in the absence of *actus reus*, not only an ethically dangerous step towards criminalising thought, but also very difficult in practice to achieve a prosecution in our legal system. The inclusion of "reasonable belief" and a decision by "ministers" likely pre-empts this legal challenge, by having the Home Secretary make judgements rather than the traditional criminal justice system.

Secondly, we risk trying to legislate our way out of the extremist mess our country faces, when we should instead be investing in non-legislative measures to tackle the causes of extremism rather than its symptoms. We have all agreed that "a poisonous ideology" is the root cause and radicalisation is the biggest challenge, yet these measures tackle neither, only serving to disrupt its symptoms – hate preachers on campuses or extremist propaganda disseminators online, for example. A third issue is that we must get beyond simply whacking whichever mole is perceived to be the current nature of the threat. Islamic extremism is like electricity, always seeking to take the path of least resistance. Rather than merely disrupting paths and wasting resources trying to catch up with extremists, we must cut this off at source, and do so without negatively altering the fabric of our nation. The fact that we cannot take water bottles onto planes now is a direct result of the foiled 2006 transatlantic aircraft plot. This will have disrupted future plots and may indeed have made us safer in the last decade, and has done so without clamping down on civil liberties, for carrying water on a plane is not a human right.

But the introduction of Terrorism Investigation and Prevention Measures (TPIMs), which curtail an individual's freedom of movement and freedom of expression without being proven guilty at a fair and public hearing, is problematic because of the failure to uphold the three aforementioned human rights. The measures now being proposed will also fail to strike the correct balance between national security and civil liberties; between counter-terrorism and human rights. Our legislation must be valued and not tinkered with every six months to catch up with the extremist tactics du jour. On top of all of this, it is plain to see that we don't have the capacity within our security services and police force to monitor even more people and enforce an ever-widening set of laws. We increasingly see extremists on the fringes of groups carrying out terrorism-related offences, rather than those at the centre. The new measures might disrupt the communication between those at the centre and those on the fringe, but what we really must do is understand why people are attracted to such groups and their ideologies, and prevent people being vulnerable to radicalisation.

We need a broader, smarter and better financed counter-extremism strategy that engages communities and mobilises civil society to understand and comprehensively challenge the Islamic ideology and refute extremist narratives. The strategy must develop resilience in our institutions and among our young people so we can address their grievances though an alternative secular, democratic, liberal lens. It must also build capacity within our civil society by engaging the private sector and training front line workers, as this is not simply an issue for the state but for all of us. These things combined will truly tackle radicalisation.

18. What is the main conclusion of the article?
A. The Government's proposed measures will be ineffective in tackling radicalisation
B. The Government must pursue a more complex strategy to counter extremism that is flexible enough to change according to the more pressing threat
C. The Government needs to be one step ahead of the terrorists
D. Terrorism can only truly be beaten if the problem is addressed at the source, with measures in place to challenge extremist ideologies in the community
E. The Government's proposed measure would be incompatible with our right to freedom of expression

19. Which of the following is not addressed as a potential problem resulting from the expansion of the law on terrorism?
A. Damaging freedom of expression
B. Extremists moving to the fringes of groups
C. Not confronting a 'poisonous ideology'
D. Difficulties in implementation (lack of resources)
E. Paternalism being taken too far

20. Why does the author use the example of Islamic extremism being 'like electricity'?
A. To scare the reader by emphasising the depth of extremist infiltration of society
B. To explain how information is shared amongst extremist groups
C. To emphasise how flexible extremists can be at finding new methods to spread their message
D. To support his point that the government will fail if they continue to address each new issues as it arises
E. To explain why the new measures will not work

21. Which of the following is a statement of fact, rather than an assertion of opinion?
A. 'the proposals are to include "extremism disruption orders", which would not criminalise the act of hate speech'
B. 'it is plain to see that we don't have the capacity'
C. 'We need a broader, smarter and better financed counter extremism strategy'
D. 'the inclusion of "reasonable belief" and a decision by "ministers" likely pre-empts this legal challenge'
E. 'we should instead be investing in non-legislative measures to tackle the causes of extremism rather than its symptoms'

### 7. Does Death Penalty Save Lives? A New Debate

Adapted from 2007 © Adam Liptak for The New York Times (edited)

For the first time in a generation, the question of whether the death penalty deters murders has captured the attention of scholars in law and economics, setting off an intense new debate about one of the central justifications for capital punishment. According to roughly a dozen recent studies, executions save lives. For each inmate put to death, the studies say, 3 to 18 murders are prevented. "I personally am opposed to the death penalty," said H. Naci Mocan, an author of a study finding that each execution saves five lives. "But my research shows that there is a deterrent effect." The studies have been the subject of sharp criticism, much of it from legal scholars who say that the theories of economists do not apply to the violent world of crime and punishment. Critics of the studies say they are based on faulty premises, insufficient data and flawed methodologies.

The death penalty "is applied so rarely that the number of homicides it can plausibly have caused or deterred cannot reliably be disentangled from the large year-to-year changes in the homicide rate caused by other factors," John J. Donohue III, a law professor at Yale with a doctorate in economics. But the studies have started to reshape the debate over capital punishment and to influence prominent legal scholars. "The evidence on whether it has a significant deterrent effect seems sufficiently plausible that the moral issue becomes a difficult one," said Cass R. Sunstein, a law professor at the University of Chicago who has frequently taken liberal positions. "I did shift from being against the death penalty to thinking that if it has a significant deterrent effect it's probably justified."

Professor Sunstein and Adrian Vermeule, a law professor at Harvard, wrote in their own Stanford Law Review article that "Capital punishment may well save lives. Those who object to capital punishment, and who do so in the name of protecting life, must come to terms with the possibility that the failure to inflict capital punishment will fail to protect life." There is also a classic economics question lurking in the background, Professor Wolfers said. "Capital punishment is very expensive," he said, "so if you choose to spend money on capital punishment you are choosing not to spend it somewhere else, like policing." A single capital litigation can cost more than $1 million. It is at least possible that devoting that money to crime prevention would prevent more murders than whatever number, if any, an execution would deter.

The available data is admittedly thin, mostly because there are so few executions. In 2003, for instance, there were more than 16,000 homicides but only 153 death sentences and 65 executions. "It seems unlikely," Professor Donohue and Professor Wolfers concluded in their Stanford article, "that any study based only on recent U.S. data can find a reliable link between homicide and execution rates." The two professors offered one particularly compelling comparison. Canada has executed no one since 1962. Yet the murder rates in the United States and Canada have moved in close parallel since then, including before, during and after the four-year death penalty moratorium in the United States in the 1970s.

"Deterrence cannot be achieved with a half-hearted execution program," Professor Shepherd of Emory wrote in the Michigan Law Review in 2005. She found a deterrent effect in only those states that executed at least nine people between 1977 and 1996.

Professor Wolfers said the answer to the question of whether the death penalty deterred was "not unknowable in the abstract," given enough data. "If I was allowed 1,000 executions and 1,000 exonerations, and I was allowed to do it in a random, focused way," he said, "I could probably give you an answer."

22. Which of the following, if true, would be most compelling in support of the idea that the death penalty does save lives?

A. A study concludes that 80% of murderers do not understand the personal implications of their actions
B. A larger study was undertaken showing a higher degree of correlation between dropping murder rates and death sentences
C. A study determines that 70% of murders are premeditated
D. The comparison between Canada and the United States is proven to have been falsified
E. A survey was circulated to hundreds of thousands of free citizens. 59% of them suggested that they would be less likely to kill someone if they knew that there was a possibility they might be executed as a result

23. What can be inferred from Professor Sunstein and Adrian Wolfers' discussion in the Stanford article?

A. Choosing not to inflict capital punishment may not result in lives being saved
B. It is unlikely that those who oppose the death penalty will change their minds following this research
C. Those who support capital punishment will be troubled by this research
D. The authors believe that those who oppose capital punishment may do so without much thought as to why
E. Murderers on death row will want this information to be circulated

24. Why does the author use the word 'admittedly' in the statement 'The available data is *admittedly* thin, mostly because there are so few executions'

A. He wants to show that the studies are worthy of discussion, despite not having many participants
B. He wants to emphasise what his personal opinion is, whilst taking notice of the opposition
C. He knows he must share the opposition to his own argument, but wants to do so whilst maintaining his own opinion
D. He wants to show reluctance in sharing the opposition argument as he knows it has merit and could weaken his point
E. All of the above

25. What, on balance, does the article suggest is the main reason these studies have attracted criticism?

A. Economic theories do not apply in real-life situations
B. The method under which these conclusions were drawn was flawed
C. The studies consider capital punishment in abstraction and do not compare with other options such as crime prevention
D. The authors who have undertaken them have been the subject of professional criticism
E. There is not enough available data to conclude whether or not capital punishment has a deterrent effect

**8. The answer to Britain's economic woes? Make pensioners work longer**

Adapted from 2015 © Lauren Davidson and Peter Spence for The Telegraph (edited)

The UK could have a stronger economy if it encouraged its older population to stay in the workforce, a new report has found.

The World Economic Forum has ranked the UK 19th out of 124 nations on an index that measures how well countries "nurture, develop and deploy their great asset – its people" with a focus on education, skills and employment. With an overall score of 79.07%, the UK capitalises on its population better than such economic powerhouses as Germany and China, but fails to match up to many of its peers including the US, France and Japan.

The UK slipped down the index because of its weak score in the 65 and over age group, coming 47th in this category, dragged down by a particularly low labour force participation rate of 10.2% that saw 86 other countries fare better than the UK.

An unemployment rate of 3.7%, ranking the UK just 38th on the global index, suggests that the silver population is lacking in job opportunities for those who want to work.

"There's always a lot of focus on youth unemployment and disillusionment which has a long term effect on a country," Saadia Zahidi, head of the WEF's Employment, Skills and Human Capital Initiative and co-author of the report, said. "But the same issue applies to women and older workers as well."

The average age of retirement in the UK is 64.7 for men and 63.1 for women, but ministers believe it should be higher to prevent healthcare and pensions crises caused by the ageing population and have previously indicated that they would like the average retirement age to rise by as much as six months every year. A recent report from the Department for Work and Pensions found that an extra £25bn would be generated if just half of the older people seeking work were given jobs.

"Talent, not capital, will be the key factor linking innovation, competitiveness and growth in the 21st century," said Klaus Schwab, founder and executive chairman of the World Economic Forum. "To make any of the changes necessary to unlock the world's latent talent – and hence its growth potential – we must look beyond campaign cycles and quarterly reports."

Finland topped the ranking, which the WEF claims is the only global report of its kind, with a score of 85.78%, followed by Norway and Switzerland. Canada, which is the world's best at maximising its 15 to 24 age group, took fourth place, while Japan rounded out the top five thanks to its chart-topping deployment of people over the age of 65.

Julio A. Portalatin, president and chief executive officer of Mercer, which produced the report in collaboration with the WEF, called the Index "a critical tool for global employers".
He said: "It allows them to determine the most pressing issues impacting talent availability and suitability around the world today and identify those issues that have the potential to impact business success in the future – invaluable insight for guiding the allocation of workforce development and investments."

26. Which of the following is an assumption that the authors have made?

A. The UK focuses too much attention on youth unemployment
B. Most over 65's currently out of work would like to work if the opportunity arose for them to do so
C. The figures suggest that there are not enough job opportunities available for over 65's (as opposed to over 65's just not wanting to work)
D. Finland, Norway and Switzerland properly utilise their ageing population
E. Talent is more important than capital to ensure a stable economy

27. Which of the following most accurately describes the index identified in the extract?

A. A critical tool for global employers
B. An indication of how efficiently countries utilise their population within the economy
C. A method by which a country's education, skills and employment are evaluated
D. A source of insight to help allocation of investment in countries' workforces
E. A set of figures which shows how successful countries are at providing opportunities for their ageing workforce

28. What does Klaus Schwab imply in his discussion of the value of talent?

A. Campaign cycles and quarterly reports are an inefficient use of time
B. Talent is more important than capital
C. Those who look only to campaign systoles and quarterly reports are naive
D. The world will not be able to meet its potential for growth unless unutilised talent is eliminated
E. Quarterly reports are useless

## 9. Deadly Frog Fungus Pops Up in Madagascar, An Amphibian Wonderland

Adapted from 2015 © Jane Lee for National Geographic

*Madagascar has been spared the scourge of the chytrid fungus, until recently.*

Madagascar is home to a mind-boggling array of frogs, 99 percent of which are found nowhere else in the world. But a study released Thursday finds the island nation now also hosts the greatest threat to amphibian biodiversity in modern times—the chytrid fungus.

As many as 7 percent of the world's amphibian species live only in Madagascar, says Molly Bletz, a researcher at the Braunschweig University of Technology in Germany. Chytrid is responsible for the decline or extinction of hundreds of amphibian species around the world. One forest in Panama lost 30 amphibian species to the fungus in about a year, according to a 2010 study.

*Why It Matters*

Researchers had thought Madagascar was chytrid-free. A 2014 study found chytrid on Madagascar frogs shipped to the U.S. for the pet trade, but researchers weren't sure whether the animals were contaminated en route or infected in Madagascar.

But a new study in the journal *Scientific Reports* finds that chytrid is present in multiple Madagascar frog species. Bletz and colleagues examined skin swabs and tissue samples from 4,155 amphibians tested for chytrid from 2005 to 2014. They found, to their surprise, that the fungus began to appear on frogs starting in 2010. What they haven't found yet is sick frogs. "It could mean we just caught it very early," Bletz says, or it's possible the chytrid strain in Madagascar isn't very lethal.

*The Big Picture*

"It's the best worst-case scenario," says Jonathan Kolby, a researcher at Queensland's James Cook University, who was not involved in the study. "[Chytrid] is there, but the frogs aren't dying right now."

Scientists need to figure out where the chytrid came from, though, he says. If it was introduced, scientists need to know how it got into the country and how they can prevent another introduction. "Because next time, it could be a strain that's supervirulent," says Kolby, a National Geographic grantee.

*What's Next*

Meanwhile, experts are working on a multipronged response to the threat. Bletz is working on a possible preventive treatment using frog skin bacteria that may fight off the fungal invader. Other groups around the world—such as in Panama—are setting up breeding facilities for especially vulnerable amphibians just in case, while others in places including Madagascar and Panama are working on long-term amphibian monitoring efforts.

29. Which of the following best describes Jonathan Kolby's approach to the situation in Madagascar?

A. Optimistic
B. Cautiously Relieved
C. Pessimistic
D. Concerned
E. Prudent

30. What can be inferred about chytrid from the article?

A. It takes some time to kill frogs
B. It is supervirulent
C. It is a fungus
D. It is harmful to all amphibians
E. It is not harmful to frogs

31. Which of the following are mentioned in the article as being worked on as a response to the threat?

A. A prophylactic treatment
B. Developing reproduction centres for vulnerable Madagascan frogs
C. Improving observation procedures
D. B and C
E. A and C

### 10. Why presume a child raised by a gay couple will be emotionally damaged?

Adapted from 2015 © Eleanor Morgan for The Guardian (edited)

In the summer holidays before my 14th birthday, my mum told my siblings and I that my dad would be leaving because she'd begun a relationship with her friend – a woman. I remember the scene clearly; Mum sat on the bottom bunk and told us, calmly, what had happened and what was going to happen now. We cried. We shouted. I ran out of the room screaming at her while my sister gingerly cuddled her.

It's gut-wrenching, whatever the circumstances, to watch your dad pack his things into sports bags and leave the family home. In a fit of utterly emo teenageness I made myself listen to John Lennon's Jealous Guy again and again on my Discman until I could cry out my frustration. I laugh now when I remember writing "I was shivering inside" at the back of my school organiser. But I was. My teenage heart hurt, too, when we'd spend weekends at his temporary bedsit in Chigwell and wake up in the night on our blow-up lilo to see him looking at us with watery eyes. But aren't these experiences par for the course of any parental separation? I had fears and prejudices relating to my mum's new relationship: what will I tell my friends? (I didn't, for ages.) What will the neighbours think? What will people think if we go on holiday together? But looking back, despite my desperate sadness, the gender shift in the house was never really a black cloud. It was the idea of someone else full stop.

This is partly why I have found pieces like Hetty Baynes' in the Daily Mail, detailing how being raised by two mums "ruined her life", and this open letter-style piece that says "Dear gay community: your kids are hurting" so uncomfortable. The layers of irresponsibility in the Daily Mail article are myriad, and while Baynes may well be damaged by her upbringing and feel justified speaking about her family in such a way, the poison dart for me is the line, "So how to explain the bizarre construct which passed for my family?"

When adults are emotionally neglectful of children – as Baynes says her mother and her female lover were of her – it can be harmful and you can end up firing your blame in all sorts of directions. But is being emotionally neglectful a gender-specific thing? I don't think so. If a woman isn't considering the needs of her daughter properly, it's not because she's gay or bisexual – these things are multi-causal and cannot, must not, be hung on the idea of gayness somehow poisoning the ability for normal love and rationality.

I had some really horrible times in the house we shared with my mum's partner, her two children and my brother and sister. Two women, five children. There were, as Baynes experienced, lots of people vying for space and prominence. There was resentment, jealousy and incidents that did cause long-term damage, but they happened as a result of two families being thrown into a building together. When I finally told my friends at school (the excuses for not having them round were wearing thin) the ones whose parents had separated and who had new family dynamics all spoke of similar problems. My problems were not unique because my step-parent figure was a woman.

I experience a crisis whenever I read any of these pieces because for every encouraging study that says the children of gay parents are doing all right we have these singular accounts of pain and turmoil attributed to homosexuality. Everyone is entitled to portray their pain how they wish, but at what point do we say, as a society, that using a public platform to make gross generalisations like Baynes does when she says "roll the dice … you are taking a chance with an innocent life" is harmful?

Although perhaps not to the same degree, I feel the same way today as I did when Dolce and Gabbana made their hateful, dehumanising remarks about "synthetic" IVF children in an interview, because such insular considerations, in such a visible context, are shameful. Talking about your own experience definitely falls under the free speech banner, but making blanket remarks about what is and isn't safe for a child doesn't.

As someone who was co-parented by two women for a while, and who is in a same-sex relationship and planning to have IVF to conceive a child some day, seeing these things splashed across front pages makes my insides curdle. I can laugh at the absurdity, but still know that, however much progress we've made, so many people will agree. That's what's frightening.

32. All of the following suggest disapproval except:
A. 'gross generalisations'
B. 'dehumanising remarks'
C. 'layers of irresponsibility'
D. 'gut-wrenching'
E. 'insular considerations'

33. Which of the following is true according to the author?
A. Her teenage reaction to her parents' news was unjustified
B. None of the problems the author encountered were due to having same sex guardians
C. The problems she encountered as a teenager was due to two families being pushed together in a new and difficult scenario
D. People should be able to speak freely about inferences about the wider world they can draw from their own personal experiences
E. Neglectfulness in a same-sex household is more often than not due to resentment

34. What assumption does the author make in the piece?
A. There is no scenario in which a child might be happy to see a father leave the family home
B. Resentment is a logical response to family upheaval
C. Dolce and Gabbana must feel similarly negatively towards same-sex unions
D. Had her father been around, she would have been happier
E. IVF children are less 'human' than other children

35. Why might Eleanor Morgan use the term 'poison dart' in relation to Baynes' argument?
A. To explain the importance of this particular part of Baynes' argument
B. To draw attention to the fact that Baynes' argument is flawed
C. To draw attention to the fact that Baynes is unjustified in extending her argument so far
D. To draw attention to the fact that the author finds Baynes' article distasteful
E. All of the above

## 11. Calculated risks: Will algorithms make business boring?

Adapted from 2015 © Zoe Kleinman for BBC News (edited)

Ever found yourself wondering whether your boss is human?

Haven't we all.

Now, with the march of artificial intelligence and robotics, it's becoming an increasingly valid question. Algorithms – problem solving computer programmes – are, to put it bluntly, getting much better at doing our jobs than we are. And it's not just in the tech sector where computers are becoming king, school children in South Korea are being taught English by a machine called Robosem.

But is there a danger that this brave new world is going to be a bit, well, dull?

*Movies by numbers*
The entertainment industry has already adopted an algorithmic approach to working out what we want to watch. When TV and movie streaming service Netflix decided to start commissioning its own material, it turned not to Hollywood veterans, film critics or media forecasters, but to algorithms and user data.

A data trawl of the most-watched and loved content streamed by Netflix customers revealed three key ingredients - actor Kevin Spacey, director David Fincher and political dramas produced by the BBC. So the firm commissioned a remake of 1990 BBC political thriller House of Cards, starring Kevin Spacey and directed by David Fincher. It became the first ever web series to win a prestigious Emmy award, with the first series receiving nine nominations.

*High tech traders*
The finance sector is another enjoying the comparative stability of computer control.
"The trading floor was a very exciting place. Now it's more like a software company than a financial organisation," said Dr Juan Pablo Pardo-Guerra, assistant professor at the London School of Economics.

High-frequency financial trader Virtu, an electronic trader with computerised strategies, has only ever recorded one day of losses in nearly six years. On its website the firm claims that its 148 employees are the "secret sauce" of its success but its own proprietary technology is at the heart of all of its trading activity.

Financial thrill-seekers will still enjoy working in the sector, predicts Dr Pardo-Guerra – but there will definitely be a change of scene. "It's probably going to have some impact on how people relate with the market, how they find excitement in trading," he said.

And the good news is there is still room for the human touch – at least for now.

"Even within the most analytical part of finance, meetings and relationships are still quite important and that's something algorithms can't do," he added. "Some aspects can be automated – calculated or quantitative decision making. But others rely more on personal cues and networks."

Algorithms have another potentially fatal flaw – they are unable to recognise when someone is taking the tricks of the trade too far.
"People know how to manipulate prices to get more customers or make their strategies more profitable," said Dr Pardo-Guerra. "I think algorithms present challenges in terms of identifying these processes."

36. Why might the author choose to start the piece in this way?
A. To grab the audience's attention
B. To establish an informal register from the beginning
C. To immediately create a connection with the audience
D. All of the above
E. A and C

37. Which of the following is not identified in the piece as an example of artificial intelligence being used to make our lives easier?
A. Computers being used to teach children
B. Algorithms being used to predict human behaviour
C. Automating quantitative decision making
D. Strategising trading decisions
E. Algorithms being used to develop products

38. What is the main reason is it suggested that artificially intelligent beings are inferior to human beings in certain roles?
A. They cannot build relationships and trust with other humans
B. They are unable to identify and respond to uniquely human reactions
C. They take tricks of the trade too far
D. They cannot identify when a human is playing outside of the rules
E. Their efficacy will result in business becoming dull

39. What does the author imply in the piece about the future of business?
A. It will be computer-centric
B. It will be more efficient
C. It might become tedious
D. Humans will still be necessary for certain tasks
E. All of the above

## 12. U.S. opposes honest labelling of GMO foods

Adapted from 2010 © Ethan Huff for Natural News

The official U.S. position on genetically-modified organisms, also known as GMs or GMOs, is that there is no difference between them and natural organisms. Crafted by the U.S. Food and Drug Administration (FDA) and the Department of Agriculture (USDA), the position set forth to the Codex Alimentarius Commission on the issue goes even further to suggest that no country should be able to require mandatory GMO labelling on food items, even though science shows that GMOs act differently in the body than do natural organisms and are a threat to health.

A group of over 80 food processors, farmers and consumer organizations has sent an official letter to Michael Taylor, deputy commission at the FDA, and Kathleen Merrigan, deputy secretary of agriculture at the USDA, protesting the official U.S. position, citing the fact that it creates "significant problems" for all U.S. food producers that wish to label their products as being GMO-free.

Not only is there no mandatory labelling of products sold in the U.S. that contain GMO ingredients, but the FDA and USDA now want to prohibit the labelling of products that do not contain GMO ingredients. In other words, the FDA and USDA are trying to outlaw truth in labelling and are openly working deceive the public.

Among those opposing the draft U.S. position on GMOs are members of the Consumers Union, the Organic Trade Association, the Union of Concerned Scientists and the Centre for Food Safety.

The FDA and USDA actually had the audacity to include in the draft position that mandatory labelling of GMOs is "false, misleading [and] deceptive" because it implies that there is a difference between GMO ingredients and non-GMO ingredients.

Fortunately, science and pure common sense, which are both lacking at the FDA and USDA, indicate that GMOs are different from non-GMOs, and that the public has a right to know the types of ingredients that are in the products they buy.

Many countries already require food processors and manufacturers to label products that contain GMOs, but the FDA and USDA hope to convince the Codex Committee to outlaw this practice.

Not only are GMOs structurally different than non-GMOs, but GMOs are actually toxic. Several studies have shown that they are harmful to the body.

40. What, according to the article, is the U.S. position on GMOs?
A. They are identical to natural organisms
B. No country should be able to demand labelling on products identifying them as GMOs
C. There is no distinction between them and natural organisms
D. A and B
E. B and C

41. Which of the following would best describe the author's perception of the U.S. position?
A. Evasive
B. Obstructive
C. Impudent
D. Illogical
E. Competent

42. Which of the following, if true, would most undermine the author's position?
A. The 'several studies' he mentions in the final paragraph are proven to have been falsified
B. The U.S position is supported by 90% of MEDC countries around the globe
C. The U.S has used no scientific evidence to support its position
D. The author is proven to be a fanatical eco-terrorist
E. An article published in the New Scientist magazine supports the idea that GMOs act no differently to biological organisms

**END OF SECTION**

**YOU MUST ANSWER ONLY ONE OF THE FOLLOWING QUESTIONS**

Your answer should be a well reasoned argument, showing your understanding of the issue in question and giving the reasons for any opinions you hold.

1. Tennessee currently protects teachers who wish to teach children to explore the potential value of following creationism. Do you think that this is correct? Identify and analyse any legal problems that may arise in discussion of this law.

2. "There is a time and a place for censorship of the internet." Discuss with particular reference to the right of freedom of expression.

3. "The UK should codify its Constitution' Discuss.

4. "The general trend towards the liberalisation of marriage undermines its religious basis." Discuss this comment with reference to the idea of abolishing marriage as a legal concept.

## END OF TEST

# ANSWERS

# Answer Key

| Paper A | | Paper B | |
|---|---|---|---|
| **1** | C | **1** | B |
| **2** | D | **2** | E |
| **3** | B | **3** | C |
| **4** | C | **4** | D |
| **5** | A | **5** | A |
| **6** | D | **6** | B |
| **7** | E | **7** | B |
| **8** | A | **8** | D |
| **9** | A | **9** | E |
| **10** | B | **10** | B |
| **11** | E | **11** | C |
| **12** | C | **12** | D |
| **13** | B | **13** | A |
| **14** | D | **14** | D |
| **15** | E | **15** | B |
| **16** | A | **16** | C |
| **17** | C | **17** | A |
| **18** | C | **18** | D |
| **19** | B | **19** | B |
| **20** | D | **20** | C |
| **21** | E | **21** | A |
| **22** | D | **22** | B |
| **23** | A | **23** | D |
| **24** | E | **24** | D |
| **25** | E | **25** | E |
| **26** | E | **26** | C |
| **27** | A | **27** | B |
| **28** | C | **28** | A |
| **29** | B | **29** | B |
| **30** | D | **30** | A |
| **31** | B | **31** | E |
| **32** | A | **32** | D |
| **33** | B | **33** | C |
| **34** | E | **34** | A |
| **35** | C | **35** | C |
| **36** | A | **36** | D |
| **37** | E | **37** | B |
| **38** | D | **38** | A |
| **39** | C | **39** | B |
| **40** | E | **40** | E |
| **41** | B | **41** | C |
| **42** | C | **42** | A |

# Mock Paper Answers

## Mock Paper A: Section A

**Extract 1**

1. C is the correct answer. A is not incorrect, but is not the purpose of the comparison. B is about the importance of universality, not efficiency. D is not particularly relevant. E is correct, but it is more of a *conclusion* of the author, not the purpose behind the comparison
2. D is the correct answer. A is overtly mentioned in the first sentence. B is not really relevant to the piece, there is no implication that the younger generation will change their gaming habits on the basis of efficiency. C is explicitly mentioned in the sentence beginning 'Now, one…'. D is almost certainly correct, but is still an assumption which leads to the assertion in the first paragraph. E is irrelevant.
3. B is the correct answer. A is incorrect as it appears to be a rhetorical question. B Is most likely to be correct, as it draws parallels between the choice that the younger generation make on a day to day basis, and the choice that they often do not think about. By bringing them into the same sentence, he makes them seem equal. C is not correct, as although he does mention that younger people have lots of choices to make, he does not imply that this is overwhelming. D is correct, but the question is placed there in the context of the paragraph, which is clearly to do with other choices to be made. E, again, might be correct, but is not the reason he concludes the third paragraph in that way.

**Extract 2**

4. C is the correct answer. Although the statement says that the researchers 'propose', it is a statement of fact that this is what the researchers propose.
5. A is the correct answer. The author mentions that even though it may not help academically, it can help create a human being with good habits. This, along with the author's tendency to consider homework in a positive light implies that good habits and personality is as important as academics. B is stated outright. C, although may be inferred, is not necessarily implied by the author in the first sentence. D seems a little too far-fetched. It could be implied that busy work is not as important as other work, but it's too much of a leap to assume that she meant to imply that these teachers are *failing* their students. E, again is a step too far. It is clear form the piece that she does not agree with Kohn's work, but to conclude that it is untrustworthy would be incorrect.
6. D is the correct answer. This question refers to the second sentence of the last paragraph. A assumes that earning money is the equivalent of success. B refers to the child's mindset, which is not the same as 'methods'. C is a direct quotation from earlier in the extract, but this is not asserted as something that a child *needs*, instead it is stated as an inevitable result of children doing homework. E is irrelevant.
7. E is the correct answer. This question will be between C and E. C, however is too black and white, saying that homework is *not* helpful for academic achievement, where the article is less direct on that issue.

**Extract 3**

8. A is the correct answer. Although B is stated explicitly in the text, it is not the reason *why* the author uses the example, instead it is a question that is raised as a result of the use of the example. C cannot be correct as the author cannot emphasise a point that he has not yet made. D is incorrect, not only for the purpose of the example, but the piece altogether, as is E.
9. A is the correct answer. This is simply a question of close textual interpretation. A is identical to the first sentence of the last paragraph. B states that they are entirely different concepts, therefore not linked, which is not true. C is not true, as although the last sentence states that they must be *viewed* in the same way, the identities themselves are clearly not identical. D is not really dealt with at all, and E is not dealt with in the final paragraph (and is an incorrect interpretation of what was said previously)
10. B is the correct answer. B is taking a statement too far. The extract only suggests that the theory of tolerance is 'yet to be tested', not that tolerance has failed. A and C are stated in the piece (with different words). D is implied by the phrase beginning 'whatever else 'Britishness' might be'. E is implied with the brackets '(as we might be of 'Englishness' if we looked close enough)'.

**Extract 4**

11. 11. E is the correct answer. All of the statements could be a valid reason why the author has included the reference to certain airlines.
12. 12. C is the correct answer. This may well be a method used by retailers, but is not mentioned outright in the article. A is mentioned in relation to furniture shops. B is 'drip pricing'. D is the 'BOGOF' offers. E is in relation to the paragraph about 'price discrimination over time'
13. B is the correct answer. This question is about very close interpretation of the structure of the question. There is a difference between 'Why is it suggested that "people are not as rational as the standard economic model implies"?' and 'Why is it suggested that people are not as rational as the standard economic model implies?'. The use of a quotation in the first example implies that the question is asking why the author suggests something, or why the author uses a particular piece of evidence. However, the second structure is really asking 'Why are people not as rational as the standard economic market implies'. The inclusion of the 'is it suggested that' is only to remind the reader that the following part of the sentence is only a suggestion by Heuristics, it is not fact. Thus, the students may get confused between A and B. C is another example of the incorrect analysis of the question. D is incorrect as it is too strong a conclusion to draw. E is simply incorrect.
14. D is the correct answer. This will likely be a choice between A and D. Although the last paragraph lists some advice for consumers, this seems almost an odd addition to end the extract with. The purpose of the main extract, excluding this section, seems to be to inform the reader about the different methods used. Thus, overall, the extract seems to be informative, rather than advisory. The last paragraph is simply an addition on how to best make use of the information gained from the piece. C is also focusing on the 'advisory' last paragraph instead of the meaning of the piece as a whole. E is a homage to the title of the piece, but is not really relevant to the content.

**Extract 5**

15. E is the correct answer. A is stated. B is not an assumption. It may be implied by the author, or inferred by the reader, but an assumption is something that must be true in order for the author's argument to make sense. B does not fulfil this criterion. C, again might be implied by the author, especially for subjects that he considered irrelevant (namely homeopathy in this instance), but it is not rightly an 'assumption'. D is not relevant to the discussion of university as an investment. E is the correct answer, as, for the author's argument about university being an investment to be correct, he must admit that there are no other reasons why someone could want to go. If he admits that they are other reasons, then his argument is fatally flawed.
16. A is the correct answer. A is explicitly stated at the beginning of the penultimate paragraph. B is only correct in relation to non-Russell Group universities. C is incorrect, as the author only implies that university is too expensive for certain degrees. For D, the author never goes so far as to suggest that the degree is worthless, he just suggests that it is not worth the fees. E is incorrect, as the author suggests only that Russell Group universities are worth attending. This extends further than Oxbridge.
17. C is the correct answer. The author describes how university is beneficial for some people but not for others, his approach to this suggests that he believes the system is flawed. A is clearly an accurate description of the author's views of some parts of the university system, but not others. B is the option which may cause the most confusion for students. Whilst the passage does not seem to be supportive of the university system, it does make clear that there are certain universities that are beneficial to their attendees. Thus it is a step too far to argue that the author is not supportive of all aspects of the university system. D, again is too far in the opposite direction. The author, despite arguing that some aspects of the system are overpriced and unhelpful, does not conclude that the system as a whole is completely useless. E is clearly wrong.

**Extract 6**

18. C is the correct answer. A is pretty much irrelevant, but for the fact that money for arts could be going on emergency services, but this is a weak argument. B would strengthen the author's argument. For E, correlation does not necessarily mean causation and thus is not as strong as it would seem. C and D could both potentially weaken the author's argument, but C is more damaging to the author's perspective.
19. B is the correct answer. The use of the word 'desperate' indicates the author's aversion to Ms Miller. A is not the author's opinion, it is a direct quote from Richard Morrison. C is not about the author's distaste for Ms Miller, but a description of the quote that he has just used. D is clearly positive. E does not show any opinion either way, it is merely a statement of fact.
20. D is the correct answer. A is stated at the beginning of paragraph 4. B is stated in the sentence 'The knives appear to be out for her in government'. C is stated in the first paragraph. E is clear. D is the answer as, even though the article states that 'much of the effect of the cut might be ameliorated through the National Lottery', it does not go so far as to state that it will make up the whole of the 5% cut in the arts.
21. E is the correct answer. A is incorrect as the question is not whether they are currently widely accepted. B is the option that the students are most likely to get confused with. This is very slightly wrong, as the question is not whether the arts are more worthy than they once were (although this is mentioned earlier), the question is whether they should be get 'special treatment'. Furthermore, it is not about being 'widely accepted' it is about whether local authorities will accept the idea. C is not really relevant - there is no mention of different opinions. The 4th answer is a moral question -should it be considered a special case, instead of whether it will be considered a special case.

**Extract 7**

22. D is the correct answer. A is mentioned in the observation that 'almost all respondents grew up in a male-dominated household'. B is explicitly mentioned in that 96% of respondents were male. C is mentioned in the penultimate paragraph.
23. A is the correct answer. B is causally related, as shown in the penultimate paragraph. C is mentioned in the exploration of the respondents' male-dominated households. D a relatively obvious point shown in the first sentence of the second paragraph. E is more subtle, with reference to people being 'pushed over the edge' in response to 'all Muslims [being] treated as terrorists'. A is the only one that is not causally related, there is no mention that the domination of extremism in the media has led to more extremism.
24. E is the correct answer. A and B are both mentioned in the extract, but neither is mentioned as being 'the definitive reason'. C is not even mentioned in the extract (though it may be inferred). D seems to be correct, as none of the above are 'the definitive reason'. However, choosing this option would admit that there is, in fact, one definitive reason. This is incorrect, as the author does not, at any point state that one correlation is more important than any of the others.
25. E is the correct answer. A is mentioned in the main body of the text, but is not the conclusion of the final two paragraphs. B is not mentioned at all, so is clearly wrong. C and D are mentioned in the penultimate paragraph, but are only part of the conclusion, not the main conclusion.

**Extract 8**

26. E is the correct answer. A is incorrect, but is likely to be picked by students. The growth of online pharmacies is not the reason why the misuse of medicine is becoming more of a topic. If anything, the growth of online pharmacies has led to medicine becoming more readily available which has in turn made it more of an issue. However, there is a missing link in the analysis to say that the growth of online pharmacies had led to a large issue of misuse of medicine. For example, the growth of online pharmacies could conceivably make it less of an issue, as it could make it more difficult for people to get medicine - perhaps because it's more difficult to prove to a computer that you deserve the medicine. B, too, is not directly responsible for the rise in relevance of misuse of medicines, it is only a factor. C and D may be inferred from the first paragraph, but it is not suggested. E is directly mentioned by the Chairman.
27. A is the correct answer. A is implied by Prof. Iversen in his 'Cat and mouse' analogy, where he states that the cat 'should not withdraw in defeat'. B and D are stated explicitly and thus cannot be said to have been implied. C is implied, but not by Prof. Iversen. Though both academics discuss pregabalin, it is only Professor Hill who mentions anything to do with the drug's widespread use. E is simply incorrect.

28. C is the correct answer. This question has been added to see if the students can determine whether a new person is being referred to, or simply more information is given about an already present character in the story. For example, the first paragraph refers to 'the government's senior drug adviser' and the second to 'Professor Les Iversen', where these are in fact the same person. The correct answer is C, the three people being Professor Iversen, the Home Secretary and Professor Hill.

**Extract 9**

29. B is the correct answer. A is incorrect - the argument relies on the premise that late-term abortions are legal, not that they have recently become legalised. B is correct, as the author simply assumes that mothers have the choice to terminate the pregnancy - this statement relies on the assumption that mothers alone are legally able to make this decision. C is incorrect - the author suggests that it may be more positive to wait until a child is born to decide whether or not to terminate its life, as there will be certainty that the child has no chance of survival. This would lead to more cases of infant euthanasia, but the reason they have increased in frequency is development in ultrasound scans such that spina bifida can be detected before birth. D is explicitly stated in the description of the illness as 'lethal'. E is a difficult one - the author may be implying this, by stating that it is 'not necessarily the best course of action', or indeed the reader might have inferred it, but it is not quite correct to state that the author assumed it in his argument - an assumption is an assertion of a fact or set on facts upon which an argument is based . This is not quite the case here.
30. D is the correct answer. In this case, incorrect students will most likely pick E, as it is the least controversial statement and thus most likely to be right. But the author does more than simply state that more discussion should be had. Throughout the article he mainly suggests positive things about a narrow legalisation of the euthanasia of children, both explicitly and implicitly, by favouring the arguments and countries that have accepted it. Further, in the conclusory paragraph he states 'parents, physicians, hospitals and nation need to confront this issue'. This is one step further than simply 'discussing': the use of the word 'confront' implies that they should decide positively on the benefits of legalisation and make changes.
31. B is the correct answer. Proof of parents supporting child euthanasia suggests that even parents, whose natural instincts are to care for their children, agree that in certain circumstances child euthanasia is beneficial. Both A and D suggest that spina bifida is a bad disease, which would support the view that late-term abortion is preferential to child euthanasia and thus sit against the author's stance. A significant rise in child mortality rate, as suggested with C is also detrimental to the author's cause. At first glance E may seem to be positive, as it looks like the President of the EU supports the author's plight, but upon further inspection, the President does not actually support the legalisation of child euthanasia, but merely stated that it should be explored more. This, in itself, does not support the author's argument.
32. A is the correct answer. A is the most likely of the correct answers. B may well be true, but it does not explain why the author began with that particular sentence. C may well be true of the first paragraph, as a push (though it seems more likely that the paragraph is included to shock the reader), but not of the first sentence, as this does not inform of the methods used, merely that they are easy. D is incorrect as both A and B could be possible reasons: E is incorrect as C is not a possibility.

**Extract 10**

33. B is the correct answer. This is a 'process of elimination' type question - A is not the 'ultimate aim', it is simply the method used. C is merely an intermediate step towards meeting the ultimate aim. D is incorrect, as the article does not suggest that this method can be used to stop any other disease except for Dengue Fever. E is irrelevant. Therefore, it must be B.
34. E is the correct answer. This question is likely to be a toss-up between A and E. Both B and C basically conclude that the project will not work, but this is not terrible, as it doesn't have any negative impact, it only reduces the potential positive impact. D shows that the evidence might be wrong, but this is not as bad as it definitely being wrong. The problem that arises between A and E is that, whilst one (A) is only a potential problem, if it were to arise the consequence would be worse, whereas (E) is a definite problem, but the consequence is not as bad as would arise if A came to fruition. On balance, though, the most damning statement should be E, as the fact that it is definitely transmittable is more damning, especially given that the consequence (infertility) remains quite serious. If for example, E were 'The discovery that Wolbachia can be transmitted to, and causes the sniffles in, humans', then on balance A would be more damning.

35. C is the correct answer. A is addressed in the reference to 'transparency and proper information for the households'. This implies that without such information, people may panic. B and E are directly addressed in the phrase ending 'concerns at the time that Wolbachia could infect humans and domestic animals'. D is less obvious, but arises from the fact that the author reassures the reader that, using this process, researchers will not have to 'constantly release more contaminated insects'. This suggests that, were the process not as adequate, this may have been a problem. C is the correct answer because, though the 10 week period is mentioned in relation to the experiment in Australia, there is no suggestion that taking longer than this would be problematic.

**Extract 11**

36. A is the correct answer. For A, the 'word seriously' is mean as a descriptive word about how affected the women's liberation movement is. B brings with it images of people putting their two cents into an argument- 'weighing in' usually carries with it negative connotations. C is pretty clearly negative, with the use of the word 'sacrifice' immediately bringing negativity with it. For D, the use of 'the sort of men' again carries with it less than friendly implications. E's use of the description 'hot pink', which tends to be associated with air-headedness for the word 'equivocation' suggests that she does not believe that it is.
37. E is the correct answer. Although some of the reasons identified are more prominent than others (for example, explaining what the fonts are like is probably not the main reason that Laurie Penny uses the word) all of them are possible reasons.
38. D is the correct answer. The answer most likely to be considered sarcastic is C. However, this is not sarcasm in the true sense of the word. Sarcasm occurs when someone says the opposite of what they mean in order to draw attention to it. The archetypal example being a grumpy teenager moaning 'oh great' in response to being made to do something that they don't want to do. C is not an example of sarcasm. It is factitious and full of attitude, but is not rightly categorised as sarcastic. More accurately it would be described as litotes – understatement for comic effect.
39. C is the correct answer. A is found in the idea that feminists want to change the may people think - they 'ask men to embrace a world where they do not get extra treats. B is mentioned in the discussion of why you cannot simply 'rebrand' feminism. D is implied by the paragraph beginning 'It is not "young women today" who need to be convinced that feminism remains necessary and "relevant". E is explicitly stated: 'the ugly, unfuckable feminist'. C, on the other hand is used to describe women, specifically how the magazines want women to be - it is not directly related to feminism as a concept

**Extract 12**

40. E is the correct answer. A is clearly wrong. B, C and D are all possible interpretations of what he's said, but his main opinion is E, as he mentions the implications for further study three times.
41. B is the correct answer. A is incorrect as the dust is simply the 'petri dish' of the experiment, the surface upon which the chemical reactions take place. C is correct in that it is radioactive, but there is no mention of it being an atom. D is clearly incorrect as the first paragraph notes that Iso-propyl cyanide is 'closer to' complex organic molecules. This by its very nature means that it is not the same thing. E is incorrect as it is only 'a step closer to discovering [...] the precursors ... of amino acids.'
42. C is the correct answer. A and D are incorrect as they are stated explicitly. B is incorrect as it is not about amino acids. E is incorrect as it is never mentioned that amino acids are essential, but only a possible way to explore life outside of earth. C can be inferred by joining together the ideas that Iso-propyl cyanide is radioactive, and that Iso-propyl cyanide could be the 'origin of more complex branched molecules, such as amino acids'

**END OF SECTION**

# Mock Paper A: Section B

**1. "The government should legalise the sale of human organs" Discuss.**

**For**

- Bring it out of the black market
- Make it safer
- Could ensure that there are certain forms, stamps etc. to make sure that a good price is being paid, the person is consenting etc.
- Stop the need for human trafficking
- Bodily autonomy
    - People should be allowed to do whatever they want with their body
    - We allow this when we are *donating* organs - why not allow it into the capitalist system - it is proven to deal best with supply and demand issues
- Will allow desperate people a final option
    - Is it not better to sell organs than sell your body or descend into a life of crime

**Against**

- Why should we allow rich people with the means to pay to be able to buy organs to prolong their life, where other people may not be able to
- Is it *right* for life itself to be part of a capitalised system
- Perhaps we already do this? Allow rich people access to more experimental and expensive medicine - perhaps the question should be more about whether all humans should have access to the same medical treatment?
- Is it *right* to sell organs - should it not remain a donation based system - is making it a commodity the best way to deal with the issue of not having enough organs? Perhaps an 'opt-out' rather than 'opt in' system would create the supply required without needed to address the moral issue of whether one should be able to sell bits of themselves.
- Do we want desperate people to have to resort to selling their own organs?
    - Legalising it may make it an option for people, where it may not have been before

**2. "Developed countries have a higher obligation to combat climate change than developing countries" Discuss the extent to which you agree with this statement.**

**For**

- Higher GDP - more 'disposable income'
- Moral obligation - MEDCs have benefitted from the earth's resources more than other countries, so have more of an obligation to make it better
- MEDCs have the technology and capacity to develop new systems to prevent/slow down climate change
- LEDC's have more important things on their mind
    - See 'pyramid of needs' below - those in MEDCs have already dealt with the bottom layer, so they can deal with the second layer.
    - Those in LEDC's cannot deal with the red layer until the yellow layer is secured

**Against**

- Do they have a higher *moral* obligation? Perhaps it is true that practically the MEDC's should be targeting climate change more aggressively than LEDC's, but maybe this is because they have more resources to do it, it is not a matter of morality
- We know that one of the main causes of methane in the atmosphere is agriculture - cows mainly. These are more likely to be found in less developed countries. Thus, logic would suggest that these countries are equally responsible, if not more so than more developed countries
- It is not disputed that MEDCs should do more, but it is not because of *morals*
- It's a universal problem - it faces everyone, so should be deal with by everyone

**3. '"Putin is a serious threat to global stability" Discuss.**

**For**

- It is a rarity in this day and age for a country to simply decide that they want land back, and just go in to take it
- His intention is ultimately to undermine democracy - he is a traditional 'baddie' - power hungry
- He cannot be defeated by manpower. He is not a 'extremist organisation which can be bombed. He is in a position of power, and thus has a unique position to undermine *global* security. He is a permanent threat
- He cannot be defeated by manpower. He is not a 'extremist organisation which can be bombed. He is in a position of power, and thus has a unique position to undermine *global* security. He is a permanent threat.
- Having such a strong, power hungry person in charge of such an extensive nuclear arsenal could result in a new cold war.

**Against**

- Need to explore whether this means Putin as a person, or Russia. How much power does Putin as a person really have, in the grand scheme of things?
- People have been scared of Russia as a concept since the cold war
- Its perceived character makes Putin seem more intimidating than other, similar people
- Is it Putin himself that is causing a threat to global stability, or the world's reaction to his actions
- We need an enemy, and if it weren't Putin, it would be someone else
- The media needs a target - and politicians follow - e.g. - The Ukraine problem got 'boring', so the media moved onto Islamic state
- How often do we hear about the Ukraine now? And has the problem actually ended?

**4. "Sufferers of anorexia nervosa should be force fed" Do you agree with this statement? If so, evaluate at what point of an individual's disease this measure should be taken.**

**For**

- This is a question about consent to medical treatment - As a result of thumb we do not treat people unless they consent to the treatment. Sufferers of anorexia nervosa who do not want to eat clearly do not consent to being fed. However, there are situations when we let this happen, if someone is considered too mentally ill to be able to give a free and informed consent.
- Anorexia is a disease whose physical symptoms all stem from the person not eating. Often the only way to save someone's life is to give them food. If they do not consent, it is indicative of the fact that they are incapable of logical thought, and the state should step in to help them, despite the intrusion on their bodily autonomy.
- There is no other option - they need food, they won't eat it.

**Against**

- Bodily autonomy - anorexia is not a mental illness like other mental illnesses - they still have their faculties and should be able to refuse treatment - even if this results in death
- Force-feeding is a humiliating, horrendous thing to have to go through (it has been used as a form of torture). There must be better ways of keeping someone alive until the disease releases its grip
- Could the humiliation of being force-fed aggravate the mental disorder?

**END OF PAPER**

# Mock Paper B: Section A

**Extract 1**

1. B is the correct answer. Firstly we can discount E. A cannot be correct as the government cannot have implied anything specific about Egypt and Sudan, as these countries are not mentioned. C is too strong a sentiment to have been implied from what is a fairly neutral statement. D is most likely to be correct, but the statement cannot stretch so far as to have implied this. They only admit to having 'redressed' the balance. This cannot be taken to mean 'retaliation'. B is correct as the government states outright that Ethiopia was left bereft of access to the Nile, which suggests that they were overlooked in the past.

2. E is the correct answer. A can be discounted outright. Both C and D show passion, but do not go so far as show disdain. B is merely a statement of fact and does not suggest how the people of Ethiopia feel. E is correct as the use of the strong 'nefarious' suggests an underlying feeling of distaste.

3. C is the correct answer. The question asks why the particular phrase is used. All of the other answers refer to the context in which the phrase is used (the building of the Dam) and do not focus on the phrase itself. B and D might confuse some students, but it must be remembered that it is not asked what Mr Hossam's argument is, nor why he chose to explain that sentiment, but why he chose to use exaggeration within that particular phrase.

**Extract 2**

4. D is the correct answer. C can be discounted immediately. A and B are both conclusions reached in the article but are not the *main* conclusion of the article. E is a conclusion that might be inferred from what the author has said throughout, but is not stated within the article and thus it would be a stretch to say that it is the main conclusion. The article keeps returning to the idea of the healthy robbing the ill of their choice to i.e. - this must factor in the main conclusion of the article and so the answer must be D.
5. A is the correct answer. C and D are expressly stated in the article so there is no room to infer them. B is not even alluded to, so to infer something so extreme would be illogical. E is touched upon in the piece, however, the author states that his work regarding those who wish to die with dignity should be commended, so clearly he is rightly placed to discuss *some* aspects of the laws. The correct inference here comes from the statement 'the fact remains that he is a rich, powerful man and it is highly unlikely that his wishes would be ignored'. A logical inference here would be that others who aren't in his position might not be afforded the same luxury.

6. B is the correct answer. A will not support the author's argument as she clearly says that some people's opinion of what they think they will want if they become ill may change if and when they find themselves in that situation. C may rectify the problem that the author identifies with regards to the choices of the rich and famous being respected above the choices of the vulnerable, but this is not the main problem that the author identifies in relation to euthanasia and thus does not sufficiently answer the question posed. D is irrelevant. E might rectify a problem relating to euthanasia - the cost of keeping people alive, but definitely not the main problem that the author identifies in the extract. B would likely rectify the problem of a seriously ill person not being able to choose - since this analysis is as close to choice as is possible given the extenuating circumstances.

7. B is the correct answer. A is too far removed from the term to suggest that it is the reason she used it. C is the alternative answer most likely to be considered correct, but does not quite fit. The paragraph about how 'it's what they would have wanted' is irrelevant for those who have not yet passed. It would be illogical for the author to start the paragraph with a discussion that people should think further about what they say and then go on to argue something completely different. D cannot be correct as it suggests that the argument that follows is useless, but this is incorrect, the author in fact acknowledges that the statement is clearly true. E is incorrect as it cannot follow simply from using the term 'knee jerk', that the argument that follows is either good or bad, the author goes onto explain this herself in as many words, so cannot expect the term 'knee jerk' to make that implication. B is correct as the adjective 'knee jerk' casts doubt upon the value of the following statement, suggesting that it has not been sufficiently analysed or evaluated.

**Extract 3**

8. D is the correct answer. A is impossible to know for sure, B is an opinion, C required knowledge of how the poll was expected to go (which is not given in this piece). D is correct as 8/10 (i.e. *most*) of people who admitted to illegally downloading music also said that they bought music.

9. E is the correct answer. A is irrelevant. B, C and D make the evidence questionable. E makes the evidence simply false.

10. B is the correct answer. This is the only statement of opinion: 'need'. The others are clearly all fact. The students might get confused about E. This is still a statement of fact as there is a condition in there - not that they *will* rise, but that they can *expect* to, based on the data.

**Extract 4**

11. C is the correct answer. C is too draconian. B is incorrect as the schools are obliged to hand over the information. Both D and E are specifically referred to in the piece as not being correct, as Fairport central school district was found to be non-compliant when they only gave over the information once they had received consent. C is the correct interpretation of the law.

12. D is the correct answer. D is the only opinion that is not specifically stated in the extract. Although Tina Weishaus alludes to the idea that the strong tactics used by the recruiters are inappropriate for children, this is not *stated*. Thus, D, if anything, is inferred by the reader from Ms Weishaus' statement.

13. A is the correct answer. C is clearly incorrect. B may at first glance seem correct but the question asks what is the *purpose* of the inclusion of the *phrase*. The use of language may well add something to the piece, but we are not discussing why the author has used that particular literary device. We are discussing the reason the *phrase* is included, and as such it follows that the phrase is included to add something to the argument, namely to draw attention to it.

14. D is the correct answer. All of the answers can be inferred from the figures given. Most students clearly don't want to be approached, as 95% of them didn't consent. Given that recruitment among seniors hit 2% it is fair to assume that most of the 5% who consented must have come from the senior class, thereby suggesting that they are more likely to consent than younger students. Finally, it would be fair to assume that, given that recruitment figures were fairly high among seniors, that the current recruitment strategies work best amongst this group of students.

**Extract 5**

15. B is the correct answer. The students will probably end up with a choice between A and B. The correct answer is B because in the first paragraph the author clearly states 'our best theories of physics imply we shouldn't be here'. It cannot be said to be 'inexplicable' as clearly, given the piece, there could be an explanation. Thus, A is incorrect. C is incorrect as it is too strong a suggestion - that it 'can' be explained instead of it 'might' be explained. D is simply incorrect as our best theories *do not* explain our existence, and E is too strong a conclusion - it is never stated that no answer will ever be found.

16. C is the correct answer. This is described in the paragraph beginning 'So Vachaspati and his colleagues…'. Only C is identified and all others are clearly incorrect – this should be a fairly simple question as long as the candidates are looking in the right place.

17. A is the correct answer. The section in the final paragraph in the hyphens relates to the definition of the term 'new science', so only the description 'stuff beyond the standard model of particle physics' is the definition and as such the answer must be A. Students might be caught out by the section after the hyphens, but this refers to what is required for CP-violation to provide enough matter in the universe, not to the 'new science' definition.

**Extract 6**

18. D is the correct answer. B and C are both wrong because the author states that it is impossible to keep addressing the new approach as it arises, the issue must be addressed from the source. E is the conclusion of the first counterargument, but it would be incorrect to say that it is the conclusion of the whole article. A was a point that was made, but again was not the *main* conclusion of the article. The answer is D as the final paragraph concludes that the government must adopt a ground-up approach to tackling extremism.

19. B is the correct answer. A is addressed in the first paragraph of the author's counter argument, C is addressed in the second, D in the penultimate paragraph about over-stretched resources in the security and police departments, and E in the discussion of TPIMS. Although those on the 'fringes' of groups are mentioned as a point of concern, the author does not suggest that the measures will encourage people to move out to the fringes of groups, and as such B must be the correct answer.

20. C is the correct answer. A and B are clearly wrong. E is far too vague and cannot be the reason why the example is used. The choice will come down to C or D. D might seem like it might be correct, however, the question asks why the specific example is used. D would have been correct had the question been, 'Why does the author use an example in this instance'. However, the question asks why he uses that *specific* example, so the answer must have some reference to the example used.

21. A is the correct answer. This is the only statement of fact. B suggests that it is 'plain to see', but this is subjective. C suggests that 'we need...'. This too must be an opinion. D is identified as an opinion by the inclusion of the word 'likely'. E is probably the most obvious opinion as the word 'should' is used.

**Extract 7**

22. B is the correct answer. A would not be in support of the hypothesis. C and E would all be one stage of extraction further away than B. C shows that murders are premeditated, which gives the murdered the opportunity to contemplate the consequences of their actions. E is not particularly compelling as it only questions free citizens, who might and probably do think very differently to murderers. D would eliminate some evidence in favour of the opposition, but does not put forward any evidence of its own. Thus, B is the only option that gives conclusive results.

23. D is the correct answer. A is explicitly mentioned and thus inference is not required. B is impossible to infer as the likelihood of change is not touched upon. C and E are simply incorrect. It can be inferred from the words which the authors choose to use that they have some level of disdain for those who oppose the death penalty. It almost suggests that they oppose it for the sake of opposing it, and that they will not take kindly (as they must 'come to terms with' it) to new information disproving them.

24. D is the correct answer. A is simply incorrect which in turn means that it cannot be E. B is too vague and does not really address why he uses that specific word. In this question, most students will be choosing between C and D. C is not quite right in that it does not address the word itself - it gives the author's motivation but it does not explain why he chooses to use the word 'admittedly'. The reason he chooses this word is to portray reluctance in showing that his argument may be flawed. D addresses this directly.

25. E is the correct answer. D is irrelevant and not touched upon in the piece at all. A, B and C are all discussed in the article as flaws in the research, but on balance, the main reason that the author suggests the studies have attracted criticism is due to the invalidity of the conclusion as a result of insufficient data. This is shown clearly as the author keeps returning to this point through the piece.

**Extract 8**

26. C is the correct answer. A reader of the piece could infer from what Saadia Zahidi has said that too much attention is spent on youth unemployment, but it is certainly not an assumption made by the authors, and thus the correct answer cannot be A. B is simply incorrect and not suggested in the piece at all. D could be inferred from the fact that they top the ranking, but is definitely not assumed by the authors. E is an assumption made by Klaus Schwab, but the authors make clear that it is not their opinion and thus not their assumption.

27. B is the correct answer. A is not really a description of the index, but rather an opinion of how useful the index will be for employers. C describes the focus of the index, but not the main purpose of the index itself. D, once again, shows how useful the index may be and not what it actually is. E describes only the part of the index that is discussed in detail throughout the extract, but is not the only information contained within it. B is correct as it accurately describes the whole index, and basically is a paraphrasing of the description given in the second paragraph

28. A is the correct answer. B and D are both explicitly stated, so there is no room for implication. C takes the idea a little too far, and it cannot be said that Schwab intended to imply that people are naive based on what he said. E again is far too extreme to have been implied. A is the correct answer as he suggests that one thing is important and then goes on the state 'we must look beyond campaign cycles and quarterly reports'. This wording implies that he considers that these are not worth spending time on.

**Extract 9**

29. B is the correct answer. B is the most relevant emotion. Kolby clearly states that the situation is fairly positive at the moment, but has the potential to get worse. B is the only option that takes into account both of these feelings.

30. A is the correct answer. B is mentioned as something that may become a problem in the future, it would not be right to infer from this that chytrid is always supervirulent, in fact it is shown that in some cases (such as this) it is not. C and D are both stated in the article and thus there is no scope for them to have been inferred. E cannot be correct as it would be too far to assume that chytrid is not harmful to any frog – the article shows that in this specific case, for some unknown reason, it is not currently harmful to the frogs in Madagascar. A is the correct answer as the statement 'It could mean we just caught it very early' coupled with the knowledge that the frogs had chytrid on their skin but are not sick would logically lead to an inference that it takes a while for chytrid to make the frogs sick.

31. E is the correct answer. A and C are explicitly mentioned in the final paragraph, whereas the reproduction centres mentioned are not restricted to only Madagascan frogs.

**Extract 10**

32. D is the correct answer. This is the only answer that conveys sadness, the rest all have an element of disapproval or anger.
33. C is the correct answer. A is incorrect as she clearly goes on to say that these experiences are normal for a child going through a parental separation. B is incorrect as she mentions she had trouble deciding whether to tell her friends about her new familial situation. D is incorrect as she demonises Baynes for making 'gross generalisations'. E is not even touched upon in the piece. C is stated in the paragraph beginning 'I had some really horrible times…'

34. A is the correct answer. This is in reference to where the author states that it is 'gut-wrenching, whatever the circumstances, to watch your dad pack his things into sports bags and leave the family home'. In order to make this statement, she has assumed that there is no possible scenario in which a child would be happy to see their father leave, or else she could not have used the phrase 'whatever the circumstances'. B is incorrect as the author makes no assumptions about how logical resentment is as a response, merely states that it was experienced within her family. C is incorrect as she only refers to Dolce and Gabbana as an example of how Baynes' article made her feel and does not suggest that they believe anything other than what they have said that they think. D is incorrect as she does not compare how she was after the break up to how she was before the break up. She notes how unhappy the new family dynamics made her but does not suggest that she would have been happier had the break up not occurred. E is the opposite of what she suggests in her discussion of Dolce and Gabbana's comments on IVF children.

35. C is the correct answer. The author uses the term 'poison dart' after discussing that Baynes' may well be justified in speaking about her family in a negative way. This suggests that the term relates to only the extension of the argument to include all gay parents. Both B and D must be incorrect, as it has been shown that the author does not dislike the whole of Baynes' argument. A is incorrect as there is no question of what is the most important part of her argument, merely the most damning part.

**Extract 11**

36. D is the correct answer. All of the answers are potential reasons that the author started the piece in this way
37. B is the correct answer. A is discussed in the first section. C and D are mentioned explicitly in Dr Pardo-Guerra's discussion of AI in the financial sector. E is discussed in relation to Netflix. B is the correct answer as the article discusses how there are certain elements of human behaviour that computers simply cannot predict.
38. A is the correct answer. C is purely incorrect. E is irrelevant. D is true but cannot be said to be the main reason that AI beings are inferior. A and B are both identified in the piece as being restrictions that AI beings have, however, A is discussed first and at further length than B, and so it must be concluded that it is the main reason that the author considers that AI beings are inferior.
39. B is the correct answer. B is the only thing that is actually *implied* by the author in the piece. She notes that 'computers are becoming king', and not just in the tech sector. This is an explicit explanation of A. C is described in her rhetorical question that it may become 'dull' and D is discussed in the final section, with 'And the good news is there is still room for the human touch - at least for now'

**Extract 12**

40. E is the correct answer. The US position is given in two halves, the first in the very first sentence, and the second at the end of the paragraph. Thus, the answer must be in two parts, so either D or E. E is the correct answer as C is more similar to the wording given in the article, and is marginally different to the meaning in A so both cannot be correct.

41. C is the correct answer. The author uses the phrase 'The FDA and USDA actually had the audacity to include in the draft position…'. The use of the word audacity is the inspiration for this question. Although all of A through to D could be correct the *best* description of the author's position must be C, given that C is a synonym of the word 'audacious' as explicitly mentioned in the piece.

42. A is the correct answer. C is clearly incorrect as it argues against the statement given. D is an 'ad hominem' argument and only undermines the person, not the person's argument. B proves that there is support for the US position but does not disprove the author's position. An article may well undermine the position if backed up by evidence, but that is unknown. The publishing on an article may not address the issues the author raises and as such would not be as damning as the disproval of the studies that that author identifies.

**END OF SECTION**

# Mock Paper B: Section B

1. **Tennessee currently protects teachers who wish to teach children to explore the potential value of following creationism. Do you think that this is correct? Identify and analyse any legal problems that may arise in discussion of this law.**

**For - Point**

- Freedom of speech should apply to teachers as much as anyone else
- Students should be given the opportunity to choose between evolution and creationism.
- It is possible to teach creationism in a critical and analytical way so as to allow children to decide themselves on what they should believe.
- This in turn will help ensure that science class is about critical discussion rather than just imbibing facts - education instead of indoctrination
- The bill doesn't exclude evolution from being taught alongside creationism - it merely allows room for other theories to be taught.
- Evolution will still be taught in school, but the bill creates more options an opens up academic enquiry

**For – Counterpoint**

- This is not a simple freedom of speech issue.
- Teachers have freedom of speech in their own time, but during their working hours they have a responsibility to only teach what is relevant to their pupils.
- For example, teachers should not tell their student their own opinions on certain issues, nor should they divulge personal information - their job is to teach fact and curriculum
- Creationism can be taught in a religious context, but should be kept out of schools. Evolution should remain in science classes
- Allowing room for other 'theories' may not be an accurate portrayal of the situation. Impartiality and objectivity is something to strive towards, but creationism is not just another 'theory', it is active denial of scientific fact, and as such should not be fit within an academic environment
- Teaching purely evolution does not mean that students cannot discuss and analyse how well evolution fits with the facts - it can still give rise to a valid discursive environment.
- Furthermore, whilst we want children to be critical of what they are being taught, a large part of education is about the transferral of information and fact - it is widely accepted now that creationism does not have its roots in fact, and as such the teaching of it could confuse children.

**Against – Point**

- Teachers do not and should not have the freedom to teach whatever they want as fact
- Teachers should have to stick to a syllabus
- Children should be given a rounded education that includes knowledge and how to critically analyse that knowledge. However, including too many options in early education could confuse a child. In primary school children should be taught what science is, and then critical analysis could be incorporated at a later date.
- Children should have the right not to be misled by their teachers.
- Telling children that evolution and climate change are scientifically controversial is misleading, as there is no controversy amongst scientists
- Creationism is not science and should not be taught as such
- Science requires testability and falsifiability, which is satisfied by the theory of evolution, and not creationism

**Against – Counterpoint**

- It is never too early to teach students the value of questioning ideas and theories.
- Just taking the consensus view because it is the consensus view does not encourage students to question what they are told. We do not want students to just believe everything they are told, we want them to be critical and decide for themselves.
- Teaching children only one viewpoint could mislead them into thinking that the issue is fact and settled.
- Arguably, evolution cannot be tested either

**2. "There is a time and a place for censorship of the internet." Discuss with particular reference to the right of freedom of expression.**

**For - Point**

- It is the purpose of the government to protect its citizens from harmful sites
- Certain social media sites can be used to harm others, via cyber bullying resulting in psychological and even physical damage. More recently sinister terrorism recruitment sites have become more proactive and successful in recruiting young people. These results can and should be protected against, and the best way to do so is with censorship
- The government has a right to restrict free speech in certain circumstances
- People can express their beliefs and opinions but only where it does not impact other people's human rights. For example we restrict the expression of racial hatred.
- Protection of people from cyber harm would be a logical extension to the valid restrictions on the freedom of speech
- Other forms of media are already regulated
- Newspapers and books are subject to censorship, as are television and film. The internet is arguably more dangerous by virtue of its accessibility, so should be censored in the same way.
- Even sites that everyone thought were innocent have been used in devastating ways
- For example - rioters organising themselves via Facebook and instant messenger. London riots resulted in numerous acts of criminal damage, theft and violence towards others.
- In order to protect people the government must be able to censor sites that can be used disruptively

**For – Counterpoint**

- Social networking sites can be used malevolently, but they can also be used as a force for good. Surely we should not censor a site which has the potential to do good just because it also has the potential to have a negative impact.
- At what point do we decide that something's negative impact outweighs the potential value of its impact?
- Perhaps censorship *can* be justified but only where the site is objectively and completely harmful and thus the justification of protecting citizens can be used.
- Banning prejudice does not actually address the problem. Perhaps a better way to deal with the issues that arise on the internet, particularly those that arise on social media sites is to engage with it publicly, rather than trying to avoid it.
- The internet is different from other forms of media, in that it is a forum for free information and expression for individuals rather than corporations. Its value is in its freedom and to censor it is to change it irrevocably.
- As a proportion of the number of people who actually use these sites, those who misuse them are very small.
- The site was merely the platform and not the reason behind the disenchantment. The government will be continually chasing these people around the internet from site to site - censoring the site will not address the problem.

**Against – Point**

- Censorship is incompatible with the notion of free speech.
- It is hypocritical for a government to give people free speech and them ban certain areas of the internet.
- The internet is a free domain and should not be controlled by the government
- The internet is also international, so it is harder to identify which sites would fall under a particular country's laws and regulations
- Censorship may do more harm than good, particularly in terms of losing respect of the government
- For example, people are so dissatisfied with the dictatorial government in China that there has been growing public outrage and disenchantment with the authority of the government

**Against – Counterpoint**

- No rights are complete, there can always be a reason for restricting a right if it protects people.
- The relevant question is whether the government should restrict freedom of expression, but about how far into a person's autonomy the paternalistic state can venture under a justification of protectionism.
- Arguably however, if the information within a particular site has the capacity to harm people form a certain state, there is a reason and a justification for governmental intervention
- The government can weigh up whether the potential detriment caused by disenfranchisement is worth the benefit of protecting people online.
- For example, people don't like to pay taxes or for taxes to be raised, but the government has the power and the right to collect and raise taxes if they believe it to be for the good of the people.

**3. "The UK should codify its Constitution' Discuss.**

**For - Point**

- England will integrate better with Europe if we have a similar legal foundation.
- For protection. A codified Constitution would help prevent tyrannical leaders coming to power.
- For example, the German Constitution allows the German federal government to declare parties unconstitutional and dissolve them.
- A codified, written Constitution provides framework for a successful separation of powers. This is essential for a system of checks and balances.
- We already have a number of written document outlining different sections of our uncodified Constitution. It wouldn't take a lot to codify everything into one manageable document.

**For – Counterpoint**

- That suggests that we want to stay within Europe. There's a growing national movement of distrust of Europe and to change our constitution to suit Europe would exacerbate the problem.
- The likelihood of that happening is very small - we still have rules and regulations, they just aren't written down in one document.
- Does it though? We wouldn't necessarily be changing the way that our political system works - our executive would still be within our legislature.
- We shouldn't do something just because it might be easy - there needs to be a valid reason for doing it.
- Furthermore it may not be as easy as it seems - there are a number of elements of our constitution that are not enshrined in writing

**Against – Point**

- Codification of the Constitution eradicates the flexibility of our current uncodified Constitution.
- Sometimes extreme action can be necessary to protect people in new and changing circumstances (e.g. infringing on people's privacy rights in order to protect about new forms of terrorism). Having a codified constitution inhibits quick and efficient action.
- Too much emphasis on the letter of the law leads to an obsession with the text itself and can result in laws that don't quite fit with current popular opinion
- One danger of codifying a constitution is to give the judiciary too much power.
- One of the great things about Britain's common law system is that judges can adapt the law to suit real life situations. Codification could interfere with this and allow unelected officials to 'legislate from the bench'
- Codified constitutions are difficult to change
- A decision to codify our constitution would grant too much power to the government who is able to write it

**Against – Counterpoint**

- Part of the point of a constitution is to protect people's rights. It is important for a government to have to jump through certain hoops to legitimise their actions. If we don't have sufficient checks and balances the government could be free to abuse their powers.
- Codifying the constitution wouldn't necessarily mean more power for the judiciary. It can be written to protect the system that we already have.
- Furthermore the Constitution can always be changed if it isn't right
- Being resistant to change is the point of a constitution, it provides security.

**4. "The general trend towards the liberalisation of marriage undermines its religious basis." Discuss this comment with reference to the idea of abolishing marriage as a legal concept.**

**For - Point**

- Marriage from a religious perspective is between a man and a woman and the liberalisation of this - the rise of divorce and the legalisation of same sex marriage DOES undermine its religious basis, in that it's a union between a man and a woman in the eyes of god.
- This is not, however, necessarily a bad thing - one idea would be to abolish marriage as a legal concept - making a joint union for the purposes of taxation that would be between whoever wants to create that union. Marriage then would be left as a separate union in the eyes of god alone and not in the eyes of the law. Whatever happens to the legal union will be separated from the religious one.
- Religion and the law should be separated, especially given our aim to be a multi-cultural and multi-religious society.
- It is to be unfairly preferential to one group of people to integrate one religion with the law above all others

**For – Counterpoint**

- Why do we support relationships in the first place - partially mutual support that a long term commitment gives to someone, but also procreation, which is lost by extending it beyond heterosexual couples
- This may be the case if we were creating the law now, but Christianity is inherently connected to the English legal system by virtue of its history and fused past.

**Against – Point**

- Religious basis is changing - lots of Christians believe that the concept of marriage should be extended to fit in one with modern perceptions

**Against – Counterpoint**

- But this is only as a response to the law changing what the understanding of marriage is.
- Marriage is traditionally a religious concept and its being integrated with the law means that the religious community has lost control of what marriage is.

**END OF PAPER**

# Final Advice

## Arrive well rested, well fed and well hydrated

The LNAT is an intensive test, so make sure you're ready for it. Ensure you get a good night's sleep before the exam (there is little point cramming) and don't miss breakfast. If you're taking water into the exam then make sure you've been to the toilet before so you don't have to leave during the exam. Make sure you're well rested and fed in order to be at your best!

## Move on

If you're struggling, move on. Every question has equal weighting and there is no negative marking. In the time it takes to answer on hard question, you could gain three times the marks by answering the easier ones.

## Afterword

Remember that the route to a high score is your approach and practice. Don't fall into the trap that *"you can't prepare for the LNAT"*– this could not be further from the truth. With knowledge of the test, some useful time-saving techniques and plenty of practice you can dramatically boost your score.

Work hard, never give up and do yourself justice.

Good luck!

### About UniAdmissions

*UniAdmissions* is an educational consultancy that specialises in supporting **applications to Medical School and to Oxbridge**.

Every year, we work with hundreds of applicants and schools across the UK. From free resources to our *Ultimate Guide Books* and from intensive courses to bespoke individual tuition – with a team of **300 Expert Tutors** and a proven track record, it's easy to see why UniAdmissions is the **UK's number one admissions company**.

To find out more about our support like intensive **LNAT courses** and **LNAT tuition** check out www.uniadmissions.co.uk/bmat

# Your Free Book

Thanks for purchasing this Ultimate Guide Book. Readers like you have the power to make or break a book – hopefully you found this one useful and informative. If you have time, *UniAdmissions* would love to hear about your experiences with this book.

As thanks for your time we'll send you another ebook from our Ultimate Guide series absolutely FREE!

## How to Redeem Your Free Ebook in 3 Easy Steps

**1)** Find the book you have either on your Amazon purchase history

or your email receipt to help find the book on Amazon.

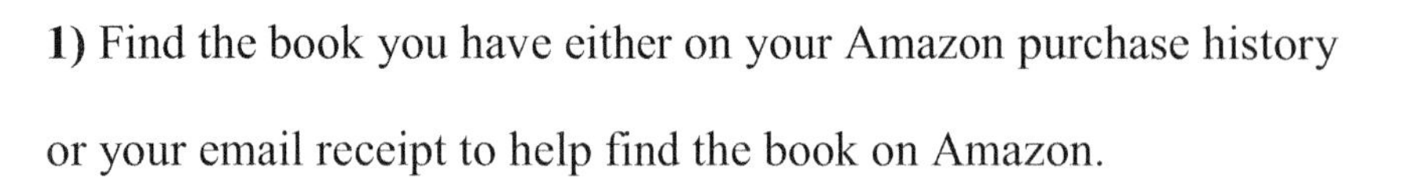

**2)** On the product page at the Customer Reviews area, click on 'Write a customer review'

Write your review and post it! Copy the review page or take a screen shot of the review you have left.

**3)** Head over to www.uniadmissions.co.uk/free-book and select your chosen free ebook! You can choose from:

- ✓ The Ultimate LNAT Guide – 400 Practice Questions
- ✓ The Ultimate Oxbridge Interview Guide
- ✓ The Ultimate UCAS Personal Statement Guide
- ✓ The Ultimate Law School Application Guide
- ✓ The Ultimate Cambridge Law Test Guide
- ✓ LNAT Mock Papers

Your ebook will then be emailed to you – it's as simple as that!

Alternatively, you can buy all the above titles at **www.uniadmissions.co.uk/our-books**

Printed in Great Britain
by Amazon